VENOM DISEASES

Publication Number 937

AMERICAN LECTURE SERIES®

A Monograph in

The BANNERSTONE DIVISION *of*
AMERICAN LECTURES IN LIVING CHEMISTRY

Editor

I. NEWTON KUGELMASS, M.D., Ph.D., Sc.D.
Consultant to the Department of Health and Hospitals
New York, New York

VENOM
DISEASES

By

SHERMAN A. MINTON JR., M.D.

Professor of Microbiology
Indiana University School of Medicine
Indianapolis, Indiana
Research Associate, Department of Herpetology
The American Museum of Natural History

CHARLES C THOMAS · PUBLISHER
Springfield · Illinois · U.S.A.

Published and Distributed Throughout the World by

CHARLES C THOMAS • PUBLISHER

Bannerstone House

301-327 East Lawrence Avenue, Springfield, Illinois, U.S.A.

© *1974, by* CHARLES C THOMAS • PUBLISHER

ISBN 0-398-03051-0

Library of Congress Catalog Card Number: 73-16496

Printed in the United States of America

N-1

Library of Congress Cataloging in Publication Data

Minton, Sherman A
　　Venom diseases.

　　(American lecture series, publication no. 937. A monograph in the Bannerstone division of American lectures in living chemistry)
　　Includes bibliographical references.
　　1. Venom—Physiological effect. 2. Poisonous animals. I. Title. [DNLM: 1. Animals, Poisonous. 2. Bites and stings. 3. Venoms. WD400 M667v 1974]
RA1255.M56　　　　615.9'42　　　　73-16496
ISBN 0-398-03051-0

For Madge, whose help and inspiration
have been invaluable

FOREWORD

Oᵁᴿ Lɪᴠɪɴɢ Cʜᴇᴍɪsᴛʀʏ Sᴇʀɪᴇs was conceived by Editor and Publisher to advance the newer knowledge of chemical medicine in the cause of clinical practice. The interdependence of chemistry and medicine is so great that physicians are turning to chemistry, and chemists to medicine in order to understand the underlying basis of life processes in health and disease. Once chemical truths, proofs, and convictions become foundations for clinical phenomena, key hybrid investigators clarify the bewildering panorama of biochemical progress for application in everyday practice, stimulation of experimental research, and extension of postgraduate instruction. Each of our monographs thus unravels the chemical mechanisms and clinical management of many diseases that have remained relatively static in the minds of medical men for three thousand years. Our new Series is charged with the *nisus élan* of chemical wisdom, supreme in choice of international authors, optimal in standards of chemical scholarship, provocative in imagination for experimental research, comprehensive in discussions of scientific medicine, and authoritative in chemical perspective of human disorders.

Dr. Minton of Indianapolis presents the biologic basis of systemic disorders caused by the bites and stings of insects, fish, mammals, and snakes. The very sharp term "venom" stirs fear of being bitten by cold-blood animals. Ancient papyri from old Egypt 1600 B.C. reveal treatments applied in Greek, Roman, and Arabic medicine. From the beginning of our record of life on earth there has been keen interest in snakes. The wily serpent offered the temptation of original sin in the Garden of Eden. The early Christians were exhorted to be wise as serpents, even though the snake is not very intelligent and the caduceus remains the modern symbol of the healing arts, using a staff intertwined with serpents.

Venoms create the most complex dilemmas devised by Nature. Indeed, the number of poisons surrounding a people is a direct function of its cultural calibre. Most of our effective drugs are poisons whose use is determined by our newer knowledge of chemical structure, toxicity, tolerance, physiological activity, and limitations. Venomous diseases are world-wide from life-threatening envenomation by snakes and scorpions at the one extreme to dangerous injuries by insects, spiders and ticks at the other, on land and in the sea. Treatment remains empirical until experimental studies in biochemistry, pharmacology and immunology further unravel the scientific basis of venom chemistry.

"The unsolved problems of how things came to be
do not enter the empirical province of objective research."

I. NEWTON KUGELMASS, M.D., PH.D., Sc.D.,
Editor

INTRODUCTION

V ENOMOUS ANIMALS and their venoms have always had a certain morbid fascination for mankind. The clinical problem of life threatening envenomation by the more dangerous snakes and scorpions, although of minor significance in the United States and Europe, remains a real concern of physicians over wide areas of the tropics. The practitioner in the United States and Europe is more likely to find himself confronted with the less spectacular but nevertheless annoying and sometimes dangerous injuries inflicted by insects, spiders and ticks. Insect stings are one of the principal causes of fatal anaphylactic shock. Spider bites, particularly those of the necrotizing type, are being recognized with increasing frequency and can be devastating in their effects. Tick paralysis can occur in almost any part of the United States. It is a potentially fatal neuropathy where a radical cure can be effected in a few seconds with nothing more sophisticated than eyebrow tweezers, but first it must be suspected. Envenomation by marine animals is increasing in frequency throughout the world as the ocean's resources are being progressively exploited, and aquatic sports grow in popularity.

Over the past quarter century the study of animal venoms has steadily moved from the sphere of the clinician to that of the scientist concerned with basic aspects of biochemistry, pharmacology and immunology. Habermann (1972) comments: "A few years ago, ingredients of venoms were dismissed as mere curiosities without relevance for general biology. Today, the view has changed. Teleologically, venomous animals practice high-level biochemical pharmacology; they have invented a series of very active, specific, pharmacologically and chemically novel drugs that may be useful in elucidating basic mechanisms of central nervous (apamin) or membrane (melittin, phospholipase A, MCD-peptide) functions."

Snake venoms have been a rich source of such compounds. The use of α-bungarotoxin in identifying cholinergic receptor protein, of a cobra venom protein for selectively depleting animals of complement, and of Russell's viper venom for studies on the mechanism of blood clotting are examples. The presence of pro-coagulant, anticoagulant, nerve growth-stimulating, and tumor cell growth-inhibiting substances in various snake venoms gives them the potential for therapeutic application. Substances such as melittin and the mast cell degranulating peptide of bee venom are useful in studying mechanisms of tissue damage, while minimine, another bee venom peptide, has a peculiar dwarfing effect on *Drosophila*. Venoms of marine invertebrates represent a virtually untapped source of substances potentially important in biology and medicine.

This book was written to bridge the gap between the multiple author, multiple volume works that generally result from international symposia and the necessarily brief treatment given to venoms and envenomation in general texts and reference works in medicine, biochemistry, and other broad fields. It has been planned as a source of information for the physician who sees envenomation largely as a clinical situation and for the student or investigator whose interest in venoms may be with their chemistry, mode of action, or evolutionary significance. It will not answer all their questions, but it may show them where some of the answers may be found and stimulate them to search for those that remain unknown.

Being venomous is an evolutionary strategy that numerous animal groups have tried with varying degrees of success. In introducing most of the major subdivisions of this book I have taken a zoologist's approach and tried to indicate how the possession of venom is of selective advantage to the animal. Venoms and the structures for administering them all were perfected long before man appeared, and his influence has, with a few exceptions, been minimal. Only one venomous species, the honey bee, has formed any particularly close association with man.

The chemical nature and biological effects of venoms can only be understood through an interdisciplinary approach involving the biochemist, immunologist, pathologist, and pharmacologist. The compiler without special competence in any of these areas can only sum up what he believes important and provide references to appropriate sources. Since there have been rapid advances in these fields in recent years, I have concentrated on the more current literature, however many of the earlier works that have contributed significantly to modern knowledge of venoms are cited.

Venom diseases are of virtually world-wide occurrence, yet their reporting is so irregular that it is difficult to estimate their incidence and public health significance. It is an accepted dictum that they are more important in the tropics and economically less developed parts of the world, but supporting evidence is surprisingly meagre. Since most of those affected are healthy children or adults in their productive years, the successful treatment and prevention of venomous wounds can represent an appreciable saving in human potential. The extent to which the psychological hazard of venomous bites and stings affects human behavior may be considerable in some circumstances but is difficult to evaluate.

The variables involved when one highly complex organism injects its toxins into another are formidable. Only those who have seen numerous cases of envenomation can appreciate the unexpected manner in which symptoms appear and develop in these diseases and how frequently the correlation between manifestations seen in experimental animals and those seen in man is less than perfect. These same variables affect the choice of therapy, and this can be a demanding test of the physician's judgement.

While this book is not intended primarily to be a guide to therapy, I realize that it will sometimes be consulted with this purpose in mind. Accordingly I have provided information on treatment of venom poisoning in sufficient detail to permit its use in practical situations. Where diverse modes of treatment are recommended, they have been evaluated as objectively as possible. The treatment of many types of envenomation is still largely empirical and symptomatic. There is a real need for physicians to report their experiences so that more effective therapies can be devised.

ACKNOWLEDGMENTS

DR. ZVONAMIR MARETIC of the Medical Center at Pula, Yugoslavia; Dr. H. Alistair Reid of the Liverpool School of Tropical Medicine; and Dr. Findlay Russell of the University of Southern California School of Medicine generously contributed advice and photographs from their wide experience with venom diseases and venomous animals. For other photographs and illustrations I am indebted to Dr. Philip W. Smith and the Illinois Natural History Survey, Dr. H. I. Rosenberg, Andrew Koukoulis, Allan Roberts, Ruth M. Sanders, Dr. Nobuo Tamiya, and John Tashjian. The helpful cooperation of the Illustration Department of Indiana University Medical Center and Mary Jane Laatz and her staff of the Indiana University Medical Center Library is gratefully acknowledged. Marie Crockett typed the final version of the manuscript.

CONTENTS

VENOM DISEASES

Chapter 1

VENOMOUS MARINE INVERTEBRATES

I. COELENTERATES

THE COELENTERATES are polymorphic aquatic animals lacking organs but possessing tentacles and a highly developed venom apparatus that is used primarily for capture of prey and secondarily for defense. There are three major groups, the Hydrozoa or hydroids in which a superficially plant-like sessile polyp stage alternates with a free-swimming medusa in most cases; the Scyphozoa or jellyfish in which the medusa stage is dominant and the polyp stage usually suppressed; and the Anthozoa, the sea anemones and corals, which have no medusa stage. Many coelenterates are colonial, and the stony corals play a major role in reef and island building in warm seas. Except for a few small hydroids, the entire phylum is marine. The major species of medical importance and their distributions are listed in Table I.

The venom apparatus of coelenterates consists of complex organoids known as nematocysts which are largely concentrated in the tentacles but may occur sparingly at other sites on the organism. Basically the nematocyst consists of a venom-filled capsul and a hollow tube often studded with spines. This is contained in a cnidoblast or supporting cell that frequently has a small projection or cnidocil. The size of the nematocyst varies from 5μ to slightly over 1 mm; the tube is tightly coiled within the capsul. Discharge of the nematocyst seems to depend upon a combination of chemical and mechanical stimuli and is triggered by the cnidocil when this structure is present. On discharge, the tube is violently everted and penetrates the tissues of the prey; at the same time, venom is discharged through its lumen. Although the content of each nematocyst is minute, the aggregate discharge from a

TABLE I.
SOME IMPORTANT VENOMOUS COELENTERATES

Physalia physalis Portuguese man-o-war; Bluebottle	Around the world in tropical and subtropical waters	Pelagic; conspicuous gas-filled float; tentacles up to 12 m long.
Millepora sp. Fire corals	Around the world in tropical waters	Typical reef inhabitants
Stephanoscyphus racemosus; Halecium sp.; *Lytocarpus* sp. Stinging Hydroids	Tropical and subtropical waters; more common in Indo-Pacific	Characteristic of reefs but may attach to pilings, rafts and other structures
Gonionemus vertens Orange-striped jellyfish	North temperate Pacific and Atlantic; Mediterranean	Reported stingings limited eastern coasts of USSR
Chironex fleckeri Australian Sea Wasp	Waters off northern Australia	Most dangerous of coelenterates
Chiropsalmus sp. Sea Wasps; Box Jellies (4 species)	Widespread in tropical and subtropical waters esp. Indo-Pacific	Cause severe stings but less dangerous than *Chironex*
Carukia barnesi Irukandji Stinger	Waters off northern Australia	Small species with 4 tentacles
Cyanea capillata Lion's Mane or Giant Jelly fish	Most of Pacific; north Atlantic	Largest of jellyfish; bell diameter up to 3 m, tentacles up to 36 m long
Chrysaora quinquecirrha Sea Nettle	Tropical and north temperate waters	Common off Atlantic coasts of U.S. Severe stinger. Tentacles to 1.2 m long
Rhizostoma pulmo Cabbagehead Jellyfish	Mostly European waters including Mediterranean	Bell diameter to 60 cm; short, heavy tentacles
Anemonia sulcata European Stinging Anemone	Eastern Atlantic and Mediterranean	Common shallow-water species.
Actinodendron plumosum Hell's Fire Anemone	Tropical Pacific	
Rhodactis howesi Matamutu	Tropical Pacific and Indian Oceans	Often eaten in Polynesia; toxic on ingestion if uncooked

large jellyfish may amount to a few cubic centimeters. Some 17 types of nematocysts have been described in coelenterates. Some function primarily as mechanical holdfasts and are not associated with venom injection. It is believed that only two types, the stenoteles and microbasic mastigophores, are significant in human envenomation. Nematocysts of a few coelenterate species are sufficiently powerful to inflict a string through rubber surgical gloves, but many are too small or weak to penetrate human skin. Detached tentacles and those attached to stranded and dying animals may continue to sting, though with diminished intensity, for several hours. Histochemical studies show that not all nematocysts are in a state of readiness for discharge at any given time; in fact occasionally a high percentage may be nonfunctional. The physiological state of the animal probably determines this and may explain why seasonal and individual variations in severity of stings by the same species may occur.

A fascinating biological phenomenon is the use of coelenterate nematocysts by other marine organisms for their own defense. Several species of delicate pelagic nudibranchs or sea slugs such as *Glaucus* and *Glaucilla* feed on the Portuguese man-of-war *(Physalia)* and transfer its nematocysts undischarged to specialized sacs in their dorsal papillae. The stinging organoids are then used by the nudibranch to defend itself against fish and larger enemies. A large number of nudibranch stings were suffered by bathers at Port Stephens, Australia in 1968 (Thompson and Bennett, 1969). The injuries were considerably less severe than those usually inflicted by *Physalia* itself.

Young of the octopus *Tremoctopus violaceus* are regularly found with fragments of *Physalia* tentacles attached in orderly fashion to the suckers of their dorsal arms. The nematocysts of these fragments are used by the octopus defensively and for capturing food. Human collectors have been painfully stung. Jones (1963) speculates that the octopus obtains the tentacle fragments by a "pickpocket type of dexterity" but also suggests that it may be immune to the toxin.

Undischarged coelenterate nematocysts have also been found in the integument of marine flatworms, and certain species of crabs carry sea anemones in their claws.

Early research workers on coelenterate venoms made use of whole tentacle extracts. Most recent workers have used material obtained from nematocysts free of tentacle tissue. Barnes (1967) has described a technique in which fresh tentacles are applied to amniotic membrane stretched across a container, and the venom discharged from the nematocysts is collected.

A growing body of evidence indicates that most of the toxicity of coelenterate venoms comes from highly labile proteins or poly-peptides. (The phenomenon of anaphylaxis was discovered during Richet's experiments with sea anemone tentacle extracts.) Lane (1960) isolated three toxic peptides from *Physalia* nematocysts. This venom also contains phospholipases A and B, proteolytic enzymes and neutral lipids (Stillway and Lane, 1971). Two types of toxin, one with lethal properties and the other a hemolysin, have been obtained from the nematocysts of the dangerous Australian jellyfish, *Chironex fleckeri* and *Chiropsalmus quadriga-tus*. The former has a molecular weight of about 150,000 and the latter about 70,000. The hemolysin of *Chiropsalmus* is more labile and displays greater activity at 5° than at 25°; there is little difference with *Chironex*. The toxins of the two species are im-munologically different; cross neutralization by antiserum was not observed (Keen, 1971). Dermonecrotic activity present in the crude venom is associated with the hemolytic fraction, at least in *Chironex* (Keen and Crone, 1969). The hemolysin is not a phospholipase, and the venom does not release histamine from the tissues (Baxter and Marr, 1969). Endean and Noble (1971) ex-tracted a toxic protein from *Chironex* tentacles from which the nematocysts have been removed. It was hemolytic but not dermone-crotic, and its relationship to the nematocyst toxin as well as its role in envenomation is uncertain. Blanquet (1972) obtained two toxic proteins from the jellyfish, *Chrysaora quinquecirrha*. The more toxic of these had a molecular weight of at least 100,000, was high in glutamic and aspartic acids, and contained no tyrosine. The toxin of the purple-tipped sea anemone *Condylactis gigantea* is a basic protein with a molecular weight of 10,000 to 15,000 (Shapiro, 1968). The toxin of another anemone, *Aiptosa pallida*, is a high molecular weight protein containing very large amounts of glutamic acid (Blanquet, 1968). A hemolytic phospholipase A has also been reported in nematocyst venom of the species (Hessinger et al., 1973) Serotonin (5-hydroxytryptamine) and quaternary ammonium compounds, particularly tetramine, have repeatedly been identified in extracts from coelenterate tentacles and other tissues but have not been found in nematocyst venom of *Chironex* or *Chrysaora*. Toxins from the fire corals *Millepora alcicornis* and *M. tenera* are high molecular weight proteins. That of *M. alcicornis* is more stable, and the two also differ in electrophoretic mobility (Middlebrook et al., 1971).

Stillway's and Lane's (1971) comment regarding the action of *Physalia* venom on experimental animals is also apropos of the action of other coelenterate venoms: "*Physalia* toxin is undoubtedly a multicomponent system. . . . It is reasonable to assume that the toxic and pharmacologic effects are due to a system of enzymes, peptides, and other factors acting in concert rather than the action of a single component." This toxin causes prompt arrest of the crustacean heart through blockade of the myoneural junction, while in the mammalian heart it causes bradycardia followed by marked electrocardiographic changes indicating abnormal conduction due to disturbed membrane permiability and distribution of electrolytes. Effects on contractility of rat intestine and crustacean gill confirm the impression that the toxin's principal effects are on membrane transport systems. When administered to intact crustaceans and fishes, the venom produces prompt paralysis that may persist for hours before death. In mice the proximate cause of death is respiratory failure; the LD_{50} is 0.75 mg per kilo (Lane, 1967; Garriott and Lane, 1969).

Nematocyst extracts of *Chironex* quickly paralyze prawns, but if sublethal doses are used, the crustaceans recover rapidly and completely. The extract produces prolonged contraction of barnacle strated muscle followed by relaxation and paralysis. The venom kills small fish but without inducing paralysis. In small mammals, injection of *Chironex* venom causes rapid death with symptoms of lethargy, ataxia, labored respirations, and convulsions. The immediate cause of death seems to be central respiratory arrest. There is an initial rise in blood pressure followed by a precipitous fall and secondary rise. There is a delay in cardiac conduction with a terminal atrioventricular block. Some degree of hemolysis is usually seen, and plasma potassium is elevated. The venom produces slow contraction of the guinea pig ileum and some release of histamine from mast cells. The mouse LD_{50} is low (0.1 ml of a 1-5000 dilution of tentacle extract) but a precise value is difficult to determine because of the extreme instability of the toxin (Endean et al., 1969; Freeman and Turner, 1969).

The pharmacological action of other jellyfish venoms such as those of *Chrysaora* and *Chiropsalmus* is not markedly different from that of *Chironex* venom. Purified fire coral toxins in doses of 1-2 mcg kill mice in a few seconds with convulsions and respiratory distress; smaller doses cause hemolysis (Wittle et al., 1971). Sea anemone toxins cause tetany and paralysis in crustaceans. The LD_{50} of *Condylactis* toxin for the crayfish is about 1 mcg per kilo (Shapiro, 1968).

Human envenomation by coelenterates is widespread in the marine environment except in very cold waters where exposure obviously is minimal. Detailed information is available for only a few areas, most notably the waters around Australia and to a lesser degree the coasts of the United States and western Europe, parts of the Mediterranean and Caribbean, and a few localities in Oceania. Jellyfish and other pelagic coelenterates such as *Physalia* may sometimes form enormous aggregations; a ship reported travelling through a school of *Physalia* for a distance of 96 miles (Fish and Cobb, 1954). Halstead (1965, Pl. XLVIII) shows a photograph of the jellyfish *Pelagia noctiluca* covering the sea surface at Heron Island on the Great Barrier Reef. Optimal conditions for inshore schooling of the dangerous sea wasps or box jellies *(Chironex* and *Chiropsalmus)* along the coast of Queensland are said to be a calm sea with incoming tide, northeast wind, and overcast sky (Halstead, 1965, p. 450, Pl. XXXIX). *Pelagia* aggregations in this region are reported to be associated with southeasterly winds, while *Physalia* aggregations along the eastern Australian coastline occur after a period of wind from the northeast (Cleland and Southcott, 1965). Another medically important

Australian jellyfish, *Carukia barnesi,* is usually encountered in shallow water during periods of north to northeast winds and a reversal of the inshore currents (Barnes, 1964).

Coelenterate envenomation also depends to a large extent upon human activities; those entering the ocean for pleasure or in pursuit of livelihood constitute the only population significantly at risk. Jellyfish stings are a genuine medical problem in Australia where up to 3000 bathers have been treated at Sydney beaches on a single weekend (Cleland and Southcott, 1965, p. 32). The sea nettle *(Chrysaora)* can be nearly as troublesome in Chesapeake Bay in summer (Burnett et al., 1968). The growing popularity of skin and scuba diving has led to an increase in the frequency of coelenterate stings by sessile species such as sea anemones and fire corals that previously were a concern only of sponge and pearl divers.

Figure 1. Portuguese-man-of-war *(Physalia physalis)* showing characteristic gas-filled float and contracted tentacles. Stranded individuals such as the one shown here can still inflict stings for a few hours. (Photo by Ruth M. Sanders.)

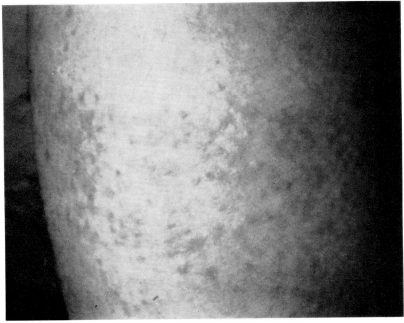

Figure 2. Maculopapular lesions following envenomation by jellyfish, *Aurelia aurita*. The broad band is characteristic of stings by this species. With many other species such as *Chrysaora,* the lesions are narrow and linear (Photo by Ruth M. Sanders.)

While most coelenterate stings are painful rather than threatening to life, more than fifty fatalities are reported from the region extending from northeastern Australia to the Malay peninsula and from Bougainville Island to Luzon in the Philippines. Where identification of the causative organism has been possible, *Chironex fleckeri* has most frequently been incriminated; a few deaths, particularly in the Philippines, have been ascribed to *Chiropsalmus quadrigatus*. There are a few fatalities reported from other regions including the coasts of the southern United States.

The usual coelenterate envenomation is manifest as a stinging or burning sensation followed by the development of a red, elevated lesion that is usually linear in the case of *Physalia* or jellyfish sting but may be blotchy or punctate in stings by hydroids and other sessile forms. In most cases it disappears in a few hours to a few days without sequelae and without systemic manifestations. Serious envenomation can result from contact with single individuals of dangerous species or multiple contacts with less dangerous species.

In several proved or suspected fatalities due to *Chironex* stings, death has occurred in 3 to 10 minutes. Both adults and children have been among the victims. There is some reason to believe contact with at least 10 meters of tentacle is necessary to produce poisoning of this degree, although the percent of nematocysts discharged is another important factor. Severe pain followed by respiratory arrest, cyanosis, and sometimes frothy sputum and convulsions are the chief manifestations. Patients who have survived long enough to reach the hospital have shown abnormal electrocardiograms, albuminuria, and leukocytosis. Autopsy findings have been limited to general visceral congestion, pulmonary edema, and skin lesions. In nonfatal cases, systemic symptoms subside quickly and completely within a few hours. Vesiculation and sometimes full thickness necrosis of the skin along the lines of tentacle contact may occur, and permanent scarring is not infrequent. Stings of this type have been reported in the Indo-Pacific region beyond the known range of *Chironex*. It is presumed that *Chiropsalmus* is the species responsible (Cleland and Southcott, 1965, pp. 43-116; Maguire, 1968).

Severe *Physalia* stings may also be accompanied by an alarming degree of systemic toxemia. Marr (1967), writing of his own experience, said that initial burning pain was soon followed by tightness of the chest, severe generalized muscular cramps, and vomiting. Similar cases have been reported by Klein and Bradshaw (1951) and Russell (1966). The serpigenous tentacle lesions may vesicate, but deeper necrosis of the skin is very rare. Cases in Atlantic and Caribbean waters seem to be more severe than those from the Indo-Pacific probably because the Atlantic form is larger and usually has several main tentacles, whereas the Indo-Pacific form usually has one. Fatalities unequivocally caused by *Physalia* toxin have not been reported, but severe stings are unquestionably dangerous. Marr, whose case is cited above, might well have drowned had he not been an experienced diver and accompanied by a companion.

Stings by jellyfish such as *Chrysaora, Cyanea,* and *Rhizostoma,* in addition to being painful, may be accompanied by cough and dyspnea with profuse mucoid discharge, muscular cramps and scrotal pain, and vomiting. Stings by *Gonionemus vertens* in Far Eastern waters may be accompanied by polyneuritis with painful paraesthesias and psychogenic manifestations such as delirium and hallucinations (Pigulevsky and Michaleff, 1969).

A unique type of jellyfish envenomation is the "Type A stinging" of Southcott (1952) or Irukandji sting of Flecker (1952). Stings of this type have been reported only from Australian and New Guinea waters and were recently proved to be due to a previously undescribed small carybdeid medusa (Barnes, 1964) which has subsequently been named *Carukia barnesi.* The initial sting is never very severe but is followed after 10 minutes to an hour by abdominal pain with vomiting, severe muscular pain and abdominal rigidity, headache, sweating, and prostration. Recovery is spontaneous within 12 to 18 hours.

Eye injuries due to coelentrate tentacles have been reported by Mitchell (1962) and Hercus (1944). Pain was severe and corneal lesions were observed, but there was no permanent impairment of vision.

Injuries by sessile coelenterates such as sea anemones and hydroids usually cause only local pain which may be severe and the development of maculopapular to vesicular lesions which may ulcerate and heal slowly. "Sponge divers' disease" described from the Mediterranean by Zervos (1934) and caused by stings of anemones attached to the sponges may cause considerable disability and even death, although the latter may be due to infection rather than envenomation. Some of DeOreo's (1946) patients who had been stung by a colonial hydroid of the genus *Halecium* suffered from cramps, chills, and diarrhea as well as dermatitis. Coral cuts and coral ulcers are a perennial plague to those whose activities bring them in contact with living coral. These injuries are not particularly painful but heal slowly and frequently suppurate. The pathogenesis is not well understood,

Figure 3. Fire Coral (probably *Millepora alcicornis*), specimen from Grand Cayman, British West Indies. This hydroid shows much variation in shape depending on the ecological situation where it occurs. Contact with it results in painful wheals. (Photo by author.)

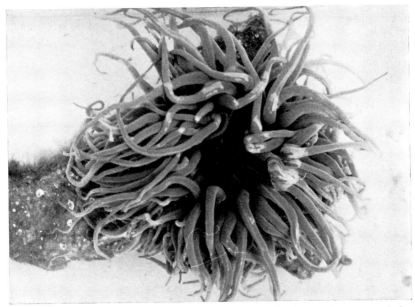

Figure 4. *Anemonia sulcata,* the most important stinging coelenterate of the Adriatic. (Photo courtesy Z. Maretic.)

but a foreign body reaction plus bacterial infection appears chiefly to blame. The possibility of sensitization to antigens in the slimy material on the surface of live coral cannot be discounted. The coral nematocysts are not involved except in the case of a few uncommon species such as *Tubastrea* and possibly *Plesiastraea.* Folklore of coral growing in wounds or body orifices is often heard in the tropics and may cause anxiety to the uninformed. Conditions such as cutaneous larva migrans or creeping eruption are occasionally attributed to coral or "coral worms."

The diagnosis in most coelenterate stings can readily be made on the basis of symptoms and history of exposure, although the exact identification of the organism is frequently impossible. The linear lesions caused by the tentacles of jellyfish and *Physalia* are highly characteristic, and clinicians with experience in treating such injuries can often be reasonably sure of the species by the appearance of the lesions. Nematocysts can frequently be recovered in identifiable form by scraping the lesions or applying

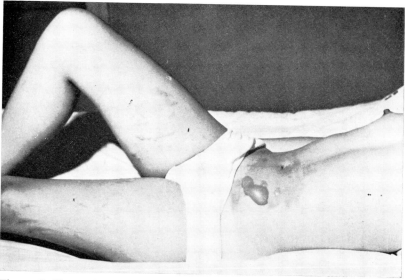

Figure 5. Hemorrhagic and vesicular lesions resulting from sting by *Anemonia sulcata*. Patient accidentally lay on anemone in shallow water. Photographed day following injury. (Photo by Z. Maretic.)

Scotch tape or similar material to them and examining the material collected under the microscope. Nematocysts may also be found in histological sections of skin, even in autopsy material preserved for a number of years (Cleland and Southcott, 1965, pp. 60-61). Nematocysts of closely related species such as *Chironex* and *Chiropsalmus* cannot always be distinguished.

Treatment of severe coelenterate stings is difficult because of the rapid progression of symptoms. Tentacles clinging to the skin should not be pulled or rubbed off, since this discharges additional nematocysts into the victim and sometimes into the rescuer. Prompt application of alcohol will inactivate the nematocysts. If alcohol is not available, salt, sugar, or dry sand may be used. After about 15 minutes the inactivated tentacles may be scraped off. If breathing stops, artificial respiration should be started at once. An open airway should be maintained and oxygen administered. Rapidly acting antihistaminics such as promethazine or diphenhydramine administered parenterally appear to be the best pharmacological antidote in severe stings. Hydrocortisone and its analogues, epinephrine, and intravenous calcium gluconate have also been reported to exert a favorable effect. Morphine or meperidine may be required to control pain. Coelenterate venoms are antigenic, and an effective antivenin has been produced against the lethal effects of *Chironex* venom but not against the dermonecrotic activity (Baxter et al., 1968). The material is expensive to produce, however, and there is the practical problem of having it available when and where it is most needed.

A wide variety of local applications have been used for milder coelenterate stings with varying degrees of success. Wasuwat (1970) reports that Thai fishermen treat jellyfish stings with a decoction of leaves of the beach morning-glory *(Ipomoea pes-caprae)*. He found this as effective as antihistaminics in relieving local symptoms of *Catostylus* sting. A mixture of picric acid, camphor, and methylated spirits is widely used at first aid stations on Australian beaches, as is an alcoholic solution of ethyl aminobenzoate (Cleland and Southcott, 1965, pp. 36-37). American practice favors use of topical antihistaminics and prednisolone. All these preparations give some degree of temporary relief, but the basic pathological process seems to be little affected. Analgesics and antispasmodics afford considerable relief in Irukandji sting and may occasionally be indicated in other types of minor coelenterate envenomation. Coral cuts usually respond well to prompt cleansing and application of a neomycin-polymyxin-bacitracin ointment.

The exclusion of stinging coelenterates from bathing areas by mechanical or chemical barriers has been suggested; however, it is often more practical and economical to exclude people at times when the animals are plentiful. Since coelenterate nematocysts have very limited penetrating power, virtually any type of protective clothing will ward off stings. Application of silicone grease to the skin will prevent stinging by *Chrysaora* (Burnett et al., 1968). A toxoid capable of immunizing persons against the stings of *Chironex* and other dangerous jellyfish can be produced; however, effective use would require immunization of large numbers of people with an expensive antigen.

II. MOLLUSKS

Among the thousands of species of shelled marine mollusks, about forty belonging to the family Conidae are known or suspected of causing envenomation in man. Commonly known as cone shells, they are mostly tropical and subtropical in distribution and inhabit relatively shallow water from the tidal zone down to a few hundred feet. They inhabit both coral and sand and are chiefly nocturnal.

The venom apparatus is associated with the proboscis which can be protruded from the anterior portion of the shell along with the tentacles and siphon. The venom is produced in a long, convoluted tubular duct whose proximal (posterior) end is attached to a large bulb which seems to serve chiefly as a reservoir of hydraulic fluid. The actual venom injection is accomplished by the hollow, barbed, chitinous radular teeth which lie in a sheath at the base of the pharynx. In most species, the teeth are 0.75 to 11 mm long. Each tooth is used only once and is charged at the base with venom from the duct just before being discharged from the tip of the proboscis.

Freshly extracted venom of *Conus magus* contains numerous spherical to sausage shaped bodies high in proteins, protein-carbohydrate complexes, and 3-indolyl derivatives (Endean and Duchemin, 1967). These are best developed in venom from the posterior portion of the duct. Proteases have been found in the venoms of some cone shells (March, 1970, 1971).

Cone shell venoms differ depending on the normal prey of the mollusk. The majority of species feed upon marine polychete worms; however, some species prey upon fish and others upon other mollusks. A few such as *Conus californicus* take more than one type of prey. Endean and Rudkin (1963, 1965) have shown that, while the basic mechanism of action of the venom is to cause paralysis of skeletal muscle, the toxicity of various species is related to their normal prey. Venoms of only six of 37 species were lethal for mice; all these species are known or suspected to feed on fish. With most of the cone shells tested, the posterior duct venom was appreciably more toxic than that from the anterior duct. Posterior duct venom of *C. magus* produced spastic paralysis and death of mice in doses of 0.2 to 0.7 mg per kilo, while anterior duct venom produced flaccid paralysis and death in doses of 4.8 to 16.6 mg per kilo. Posterior duct venoms of *C. geographus, C. catus,* and *C. stercusmuscarium* produced flaccid paralysis. Some samples of *C. striatus* venom produced spastic paralysis, while others produced flaccid paralysis. Coma followed injection of *C. tulipa* venom. The proximate cause of death in all cases appeared to be respiratory paralysis. In rat phrenic nerve-diaphragm preparations, cone shell venoms acted directly on the muscle, and the effect could be reversed by washing out the venom. Eserine and neostigmine were not effective in counteracting the paralytic effect of the venom. These venoms produced rapid paralysis in blennies and other fish. Venoms of molluscivorus cones such as *C. textile* rapidly paralyzed gastropods but had no effect on vertebrates. Venoms of vermivorus species showed activity against the polychaete *Phyllodoce,* but some also produced paralysis in blennies and local lesions in mice.

In light of their experimental results, Endean and Rudkin (1965) consider only the fish-eating cone shells dangerous to man. Considering the average size, development of the venom apparatus, amount of venom injected per sting, and degree of aggressive behavior, *C. geographus* is rated the most dangerous species followed by *C. tulipa, C. magus,* and *C. striatus.*

All cone shell stings in man have involved persons collecting shells or children who have picked them up through curiosity. Cone shells have particularly attractive patterns and shapes and are sought by both amateur and professional collectors. *C. gloriamaris* has the distinction of being the rarest and most valuable of shells. About 15 species of cone shells have been reported as causing human envenomation. Fewer than fifty cases have been reported in the medical literature; however, many others, particularly mild ones, have doubtless occurred. All reports are for the Pacific and Indian Oceans from the Kuriles to New Caledonia and from the Seychelles to Hawaii. All fatalities in recent years where the mollusk has definitely been identified have been ascribed to *C. geographus* which has an extensive range in the tropical Pacific and Indian Oceans. *C. textile* has a reputation for being danger-

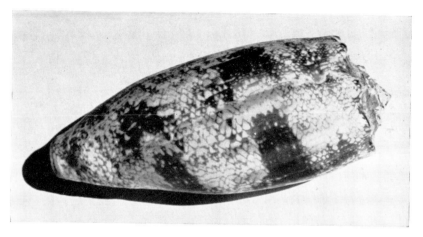

Figure 6. Two venomous cone shells, (a) *Conus geographus* and (b) *Conus striatus.* Both species are considered dangerous, and well authenticated fatalities have been ascribed to *C. geographus.* (Photos by author.)

ous, and has been implicated in serious and fatal stings, and *C. omaria* was reported to have caused a near fatal sting to a child at Manus Island in 1954. Since both these are molluscivorus cones whose venoms were found by Endean and Rudkin to be nontoxic for small mammals, either the shells were incorrectly identified or man possesses some peculiar susceptibility to their venoms. *C. tulipa* is the only other species that has been involved in serious cases of poisoning. Kohn (1963) and Cleland and Southcott (1965, pp. 199-206) have summarized most of the known instances of cone shell sting.

Pain in cone shell sting is highly variable, and in two fatal cases reported by Flecker (1936), and Rice and Halstead (1968), was not noticed by the victim. Symptoms of serious poisoning include diplopia, slurred speech, faintness, numbness, and muscular weakness progressing to paralysis. Swelling around the site of injury is slight to moderate. There is terminal coma and respiratory paralysis. Death usually occurs within six hours of the sting. Autopsy findings are few and not characteristic; cardiac dilitation and cerebral edema have been reported. Stings by less dangerous species of cones cause sharp local pain sometimes followed by swelling and numbness. Symptoms may subside in a few minutes or persist up to five days.

There is no specific treatment for cone shell sting. In severe cases, prompt use of a respirator and oxygen may be life-saving, for the systemic effects of the venom seem to be quite transient.

The only practical prevention of these stings is careful handling of live cone shells. They should be picked up by the large (posterior) end and promptly dropped into a suitable container. Several stings have been inflicted through clothing, so putting the shells in pockets is not safe. The animals should never be held with the aperture of the shell against the skin.

Cephalopods are marine mollusks characterized by having well developed eyes, 8 or 10 tentacles equipped with suckers, and a horny beak. Most species lack an external shell. Some deep-sea forms are enormous animals. Reports of giant squids and octopus attacking man are part of maritime lore, but very few have been authenticated. Human injuries by these mollusks have largely been confined to venomous bites by certain octopus species. Octopuses are widely distributed in temperate and tropical waters, most species preferring a rocky or coral habitat and relatively shallow water. They are solitary and active both by day and at night.

The posterior salivary glands of cephalopods produce a venom that helps the animal subdue its prey. The secretion is introduced into wounds made by the beak.

A protein toxin (cephalotoxin) having a paralytic effect on crustaceans but no toxicity for vertebrates was isolated by Ghiretti (1959) from the saliva of *Sepia officinalis;* it has also been found in the saliva of *Octopus vulgaris* and *O. macropus.* A similar protein has been isolated from *Eledone cirrhosa* (McDonald and Cottrell, 1972).

Eledoisin, a peptid containing 11 amino acids, has been isolated from octopuses of this genus and its composition confirmed by synthesis. It is a potent vasodilator, 50 times as active as histamine; the hypotensive dose in the dog is 0.5 to 2 mcg per kilo. It also stimulates extravascular smooth muscle and lowers the permiability of cutaneous blood vessels (Erspamer and Anastasi, 1962). Crude extracts of *Hapalochlaena* posterior salivary glands were found to inhibit neuromuscular transmission in rat diaphragm preparations and cause respiratory paralysis in the cat (Trethewie, 1965). A neurotoxin with molecular weight of less than 540 was obtained from venom of this octopus. Intravenous injection of small doses of the toxin produced almost instantaneous paralysis of mice; this effect was not counteracted by neostigmine or atropine (Sutherland et al., 1970). Other biologically active substances such as hyaluronidase, tyramine, and serotonin have been found in octopus saliva.

Octopus bites are not uncommon in parts of the world where these cephalopods are regularly taken for food, although octopus fishermen learn to handle and kill the animals in such a way that they are not often injured. These bites usually result in nothing more than local pain and swelling lasting a few hours to a few days. Flecker and Cotton (1955) reported the first fatal octopus bite in a healthy young man in northern Australia who had been handling a small, unidentified octopus. Subsequently at least two other fatal cases and several serious but nonfatal ones have been reported from Australian shores. It is believed that all have been caused by the two closely related species of blue-ringed octopus, *Hapalochlaena maculosa* and *H. lunulata,* which have a combined range around the coasts of Australia and are often common in tidal pools. These are small species, rarely exceeding an arm span of 20 cm or weight of 90 gm. When excited they develop a pattern of iridescent peacock-blue rings.

Except for one case where a small octopus became trapped under a diver's suit, all bites have resulted from handling the animal or putting it on the bare skin. The first symptom of poisoning is a giddy sensation accompanied by visual disturbances and difficulty in speaking and swallowing. This is followed by nausea, vomiting, difficulty in breathing, profound weakness, and collapse. In fatal cases, death has occurred within 90 minutes of the bite. The only significant finding at autopsy has been pulmonary congestion. In nonfatal cases, recovery is usually complete within 24 hours. One patient experienced urticaria and effusion into the joints; she had a history of allergy to shellfish (Sutherland and Lane, 1969; Edmonds, 1969).

There is no specific treatment for octopus envenomation, and the nonantigenic nature of the lethal toxin of *Hapalochlaena* presents a severe obstacle to the development of an antitoxin. As in cone shell stings, artificial respiration and oxygen is the therapy of choice in severe envenomation. Cleland and Southcott (1965, pp. 210-212) reported two cases that survived with this treatment after spontaneous respirations had ceased.

III. MISCELLANEOUS GROUPS

1. Sponges

Sponges are generally considered the most primitive phylum of multicellular animals, being without organs or well differentiated tissues. The approximately 4000 species are worldwide in distribution. A few live in fresh water, but the vast majority are marine occurring from intertidal zones to abyssal depths.

Although sponges lack specialized structures for venom injection, tissues of certain species contain toxic substances. Sponge extracts are toxic to many experimental animals, and marine animals confined with *Tedania toxicalis* usually die in less than an hour. Extracts of *Haliclona viridis* have antimicrobial activity as well. The chemistry of these substances is poorly known. They seem to be relatively stable to heat and prolonged storage.

"Red moss" dermatitis caused by contact with the sponge *Microciona prolifera* is reported to be common among oyster fishermen of the northeastern United States. The fingers are usually affected. Stiffness and redness are followed by pain and the appearance of crops of blisters. The condition may persist as long as four months (Carson and Pratt, 1943). A dermatitis lasting a few hours and followed ten days later by erythema multiforme type lesions lasting two weeks was seen in a skin-diver who handled *Tedania ignis* (Yaffee and Stargardter, 1963). Immediate burning pain, redness, and swelling, sometimes with general malaise, have been reported following contact with other *Tedania* species, *Fibulia nolitangere, Neofibularia mordens,* and *Hemocyton ferox.* Such cases are known from the West Indies, California, and Australia (Halstead, 1965, pp. 271-296). A severe itching dermatitis lasting up to two weeks but with minimal or no local lesions was seen in eight persons who handled a large marine sponge of unknown species collected at a depth of about thirty feet in St. Vincent Gulf, South Australia (Cleland and Southcott, 1965, pp. 8-11). Dermatitis from contact with freshwater sponges has been reported in Europe and Australia.

The lesions caused by sponges apparently result from mechanical irritation by spicules of the exoskeleton plus toxins in the tissue. Hypersensitivity may play a part. Yaffee and Stargardter's patient developed an anaphylactoid reaction following application of 1 per cent sponge extract.

There is no specific treatment for sponge dermatitis. Weak acid solutions such as vinegar are said to give better relief from immediate reactions than antihistaminics or corticoids; the latter are indicated in delayed reactions.

2. Annelids

Annelids are numerous in the sea inhabiting many niches and playing an important role in marine ecosystems. Several of the polychaete worms known as bristle worms or sea mice *(Chloeia, Eurythoe,* and *Hermodice)* have spines that in the case of

Eurythoe and *Hermodice* are retractile and filled with a venomous fluid of unknown composition. Contact with these spines produces burning or nettling pain, swelling, and erythema followed by numbness and itching that may last up to a few weeks. Gangrene, possibly from secondary infection, has been reported. These worms have a circumtropical distribution being particularly common in Malaya, the Great Barrier Reef region of Australia, and the West Indies. They are secretive and usually are found around rock or coral in shallow water.

The blood worm *(Glycera dibranchiata),* found along the Atlantic coast of the U.S. and Canada and commonly taken for bait, has a protrusile proboscis tipped by four fangs connected to venom glands. Nothing is known of the chemistry or action of the venom. Its bite is followed by pain and swelling lasting about 48 hours. Some of the giant polychaete worms such as *Eunice aphroditois* which reaches a length of 1.5 meters can inflict painful bites, but the presence of venom has not been demonstrated.

Venomous or irritant annelid bristles can best be removed from the skin by applying adhesive tape to the affected area and then pulling it away. This should be followed by application of an antipruritic and antiinflammatory lotion or ointment.

3. Echinoderms

Echinoderms are an entirely marine phylum. While the larvae are planktonic, the adults with a few exceptions are bottom dwellers occurring from the intertidal zone to abyssal depths. Nearly all the venomous species belong in the sea urchin group. Two types of venom apparatus are found in this group, the spines which have a purely defensive function and the pedicellariae or tube-feet which chiefly serve to immobilize prey but may also function in defense. Venom has been identified only with the secondary spines of a few species such as *Echinothrix calamaris, E. diadema,* and *Asthenosoma varium.* The toxin evidently is associated with the epithelium on the surface of the spines; fluid from the lumen of the spines is not toxic. Venom has been associated only with the globiferous type of pedicellariae which are tricuspid structures with a venom gland at the base of each section. A duct conveys the venom to the hooked and hollow tip of each valve or jaw. When a moving foreign object touches the open pedicellaria, the jaws close driving their tips into it and injecting their venom.

Alender (1967) found noradrenaline the only biologically-active substance identified in the spines of *Echinothrix* but suggested it was associated with unidentified toxic substances. Mendes et al., (1963) obtained a dialyzable substance with the pharmacological activities of acetylcholine from the pedicellariae of *Lytechinus variegatus*. Alender et al. (1965), isolated a protein toxin from pedicellariae of *Tripneustes gratilla*. It was stable over a wide range of pH but inactivated in 5 minutes at 47.5°. It evidently contained several components, the most active of which was a relatively small molecule with a sedimentation coefficient of 2.6S. The most toxic preparation contained 903 mouse LD_{50}'s per mg of protein. It had a powerful blocking effect on neuromuscular transmission in both crustacean and mammalian muscle-nerve preparations (Parnas and Russell, 1967). It also released histamine from tissues (Feigen et al., 1966).

Nearly all sea urchin injuries are caused by spine punctures and are of frequent occurrence throughout the warmer coastal regions. The degree of pain seems to vary with the species of urchin, the depth and location of the wound, and the extent to which the secondary spines are involved. Usually it is not greatly out of proportion to the mechanical injury done. The puncture wound is dark and often surrounded by a purplish ring that fades in a short time. Systemic symptoms are rare and seem to be mostly of psychogenic origin. Pedicellarial envenomation is infrequent and results from allowing the urchin to crawl on the skin or otherwise handling it. In many cases the animal either does not use its venom apparatus or the pedicellarial jaws are too weak to penetrate the skin. In addition to severe pain, pedicellarial envenomation has been accompanied by giddiness, weakness of the limbs, paralysis of the lips, tongue, and eyelids, and difficulty in breathing (Fujiwara, 1935). *Toxopneustes pileolus,* whose pedicellariae are unusually large, is feared by Japanese divers who say its sting may be fatal.

Sea urchin spines very frequently break off in the tissues and are difficult to extract because they are extremely brittle and ringed with minute spinules. Fortunately the small fragments are rapidly removed, probably by phagocytosis. Occasionally they persist in the tissue with the production of a granulomatous reaction. Secondary infection may occur but is not particularly common. There is no specific treatment for sea urchin envenomation.

The only other known venomous echinodrem is the crown-of-thorns starfish, *Acanthaster planci*. This is a large species 60 cm or more in diameter with 13-16 rays; the entire dorsum is covered with large spines. The venom-producing tissue is in the integument covering the spines. Nothing is known of the chemistry and pharmacology of the venom.

This starfish has a wide range in the central and southwest Pacific and eastern Indian Ocean. It feeds on live coral and has recently been accused of doing marked damage to reefs in parts of the Pacific. It occasionally enters the tidal zone where it may be stepped on by persons wading.

Injury by *Acanthaster* spines causes immediate severe pain with little or no swelling. This is followed in a few hours by vomiting that may be protracted and severe (Barnes and Endean, 1964). There is no specific therapy for this envenomation.

Chapter 2

VENOMOUS ARACHNIDS AND MYRIAPODS

I. SCORPIONS

SCORPIONS ARE AMONG the first of land animals with a fossil record reaching back to the Silurian, although the oldest specimens may have been partially aquatic. The group has never been particularly large and contains about 650 known living representatives. They are divided into six families with about half of all the species and all the dangerous ones in the family Buthidae. In the Old World, scorpions are found from southern Germany and Mongolia through most of Africa and Australia and in the New World from southwestern Canada to Patagonia. Their ability to fast for prolonged periods and their great powers of water conservation admirably adapt them to life in arid or semi-arid lands where the majority of species occur, although the group does not avoid humid regions entirely.

Scorpions are almost exclusively nocturnal. Desert species frequently burrow, sometimes to depths of as much as 2.4 meters. Most are crevice dwellers found under stones, bark, and other objects. A few are partially arboreal. The sexes are similar, although males of most species are more slender and a little smaller than females. The female gives birth to plump, soft-bodied young that she carries on her back during the early period of their lives. Scorpions show no other type of social behavior, but occasionally a number of individuals may aggregate in suitable microhabitat. Scorpions are predacious, feeding for the most part on insects and other terrestrial arthropods; however, large scorpions will sometimes eat small vertebrates such as lizards. Prey is usually grasped with the pinchers; it may or may not be stung to immobilize it. It is then torn and crushed with the mouthparts; only the juices and liquified tissues are ingested. Solid material is ejected as small pellets (Stahnke, 1966).

The so-called tail of the scorpion actually consists of the last six abdominal segments. The terminal one or telson bears the venom apparatus. This consists of a pair of venom glands, each enclosed in a muscular envelope. The gland tissue is made up of tall columnar cells that in most species are supported on numerous plicae. Cells containing granules that are assumed to be the toxins proper are intermixed with mucus-containing cells. Secretion is of the apocrine type. Glands of two species of *Centruroides,* one highly toxic for man and the other weakly toxic, were morphologically identical by light and electron microscopy (Keegan and Lockwood, 1971).

Scorpion venom is most readily obtained by immobilizing the animal and applying electrical stimulation to the membrane between the last two abdominal segments. Drops of venom appear at the tip of the sting and are collected in a pipette, beaker, or ampoule. An alternative and perhaps more physiological method is to allow the scorpion to sting through a sheet of parafilm and collect venom droplets from the inner surface with a pipette (Zlotkin and Shulov, 1969). Larger species of scorpion yield up to 0.022 ml of venom. Scorpion venom contains 15 to 30 per cent solids. Venom yields from some representative species of scorpions are listed in Table II. Most scorpion venoms are clear to slightly opalescent, but those of at least some diplocentrids *(Diplocentrus* sp. and *Nebo hierichonticus)* become orange to red on contact with air (Stahnke, 1970; Rosin, 1969).

From North African and Middle Eastern scorpions of the genera *Buthus, Androctonus* and *Leiurus* have been isolated a group of neurotoxins active against mice and presumably other mammals but not insects. They are low molecular weight proteins with a high content of sulfur and basic amino acids, especially lysine but without methionene. Most of them exhibit cathode mobility at pH 8.6, and they are heat labile. Similar toxins have been found in venoms of South American scorpions of the genus *Tityus* (Miranda et al., 1964a, b, c; 1966; Zlotkin et al., 1972a). The amino acid sequence of neurotoxin I from *Androctonus*

TABLE II. YIELD AND TOXICITY OF SOME
REPRESENTATIVE SCORPION VENOMS

Species	Distribution	Venom yield (mg)	Mouse LD$_{50}$ (mg/kg)[a]	Reference
Centruroides noxius	Mexico	0.075	0.150	Whittemore et al., 1963
C. suffusus	Mexico	0.155	0.315	" "
C. sculpturatus	Arizona	0.277	1.120	Stahnke and Johnson, 1967
Leiurus quinquestriatus	North Africa Middle East	0.483	0.260	Whittemore et al., 1963
			0.255	Weissman and Shulov, 1959
Androctonus australis	Northern Africa to West Pakistan	1.380	4.550	Balozet, 1955
Buthacus arenicola	Northern Africa	0.700	2.600	Balozet, 1955
Buthus occitanus	Mediterranean Region	0.290	5.750	" "
B. tamulus	India	0.400	3.000	Devi et al., 1970; Deoras and Vad, 1962
Buthotus judaicus	Near and Middle East		6.350	Weissman and Shulov, 1959
Orthochirus innesi	Near and Middle East	0.066	1.750	Shulov and Amitai, 1960
Tityus serrulatus	Brazil	0.380	1.350	Bücherl, 1953
Tityus bahiensis	Brazil	0.300	3.000	" "
Parabuthus sp.	South Africa	4.800	25-100	Grasset et al., 1946
Opistophthalmus sp.	South Africa	1.400	250-850	" "
Hadogenes troglodytes	South Africa	2.700	1500-2000	" "
Scorpio maurus	Northern Africa and Southern Europe		25-35	Sergent, 1946
Vejovis spinigerus	Southwest U. S.		4.870	Russell et al., 1968
Hadrurus arizonensis	Arizona	4.195	168	Saunders and Johnson, 1970

[a]Calculated, when necessary, from the original data assuming the average weight of a mouse to be 20 gm.

australis has been determined. It is a single polypeptide chain linked by four disulfide bridges (Rochet et al., 1970). A group of similar toxins producing contraction-paralysis and death in blow-fly larvae but without effect on mice have been isolated from this same group of venoms. One has a structure similar to that of *Androctonus* neurotoxin I. Most have moderate to fast anode mobility and are easily separated from the mouse neurotoxins by electrophoresis (Zlotkin et al., 1971a, b; 1972a). Yet another toxin active against the terrestrial crustacean *Armadillium vulgare* but not against mice or insects has been reported in the venom of *Androctonus australis* (Zlotkin et al., 1972b). Watt (1964) found the mouse lethal component of Arizona *Centruroides sculpturatus* venom is slowly dialyzable, not inactivated by pro-teolytic enzymes, and low in aromatic amino acids. Further studies (Watt and McIntosh, 1972) indicate this venom contains four neurotoxins with molecular weights of 7000 to 9000. Toxicity is dependent on the presence of disulfide bridges and lysine. Venom of the related *C. limpidus tecomanus* has moderately high contraction-paralysis activity against blowfly larvae (Zlotkin et al., 1971a). These *Centruroides* toxins appear to be similar to those of the Afro-Asian scorpions in pharmacological activity, but there are probably antigenic differences.

Adam and Weiss (1959) found high concentrations of sero-tonin in the venom of *Leiurus quinquestriatus* and much smaller amounts in *Buthotus minax* and *Vejovis* venoms. Scorpion venoms are poor in enzymes. Master et al., (1963) reported protease in the venoms of two Indian scorpions, *Buthus tamulus* and *Palmne-ous gravimanus*. The former also contained phosphodiesterase and the latter 5'-nucleotidase. Balozet (1955) reported lecithinase activity in *Scorpio maurus* and *Buthacus arenicola* venoms. Russell et al., (1968) demonstrated cholinesterase activity in *Vejovis spinigerus* venom but did not detect phosphodiesterase, protease, amylase, or L-amino acid oxidase. An acetylcholinesterase was re-ported in venom of *Hadrurus arizonensis* (Saunders and Johnson, 1970).

In mice and rats, venoms of *Buthus, Buthotus,* and *Androcton-us* cause irritability, salivation, spasticity, gasping, labored respiration, and convulsions. With large doses, the animals may die almost immediately. Secretion of reddish tears is seen in rats stung by *Orthochirus melanurus* and *Buthotus doriae*. In anaesthetized cats, *Centruroides sculpturatus* venom causes mydriasis, salivation, distention of the stomach, decreased intestinal motility, hypertension, and respiratory arrest (Patterson, 1960). *Leiurus* venom in dogs causes an increase in systemic and pulmonary blood pressure and augmented myocardial contractility accompanied by electrocardiographic abnormalities that include sinus tachycardia, ectopic beats, and abnormal Q waves. Blood glucose, lipids, and catecholamines are elevated. Focal myocardial necrosis, cellular infiltration, and fat droplet deposition in cardiac muscle fibers were seen at autopsy (Yarom, 1970). *Centruroides sculpturatus* venom was extremely potent in blocking nerve impulse transmission at the neuromuscular junction in both crustacean and mammalian preparations; four other North American scorpion venoms showed a moderate to weak blocking effect with crustacean preparations but not with mammalian (Parnas and Russell, 1967). Although scorpion venoms generally have little effect on the vasculature and clotting mechanism, Devi et al. (1970) reported disseminated intravascular clotting with increased prothrombin time, hypofibrinogenemia, thrombocytopenia, and widespread hemorrhages following subcutaneous injection of *Buthus tamulus* venom into dogs. Venoms of *Nebo hierichonticus, Hadogenes troglodytes* and *Opisthophthalmus sp.* showed only a weak neurotoxic effect in mammals but were strongly hemorrhagic and necrotizing (Grasset et al., 1946; Rosin, 1969). Venoms of a few species such as *Scorpio maurus* do not appear to be highly toxic for either mammals or insects.

Scorpions are immune to venom of their own species and probably to those of closely related species; however, *Leiurus* is often killed by stings of *Nebo hierichonticus* (Rosin, 1969). Scorpion haemolymph has considerable neutralizing power against scorpion venom (Shulov, 1955). This does not increase with immunization, and an antibody response similar to that of vertebrates is apparently not involved (Shulov et al., 1973). Wheeling and Keegan (1972) found tarantulas *(Aphonopelma smithi*) survived stings by *Centruroides limpidus;* however, the sting deterred the spider from pressing home an attack on the scorpion. Tarantula hemolymph did not abolish the lethal effect of scorpion venom for mice, but survival time was prolonged.

Toxicity of some representative scorpion venoms for mice is given in Table II. Toxicity for man may show marked variation among closely related and morphologically very similar species. In the genus *Centruroides, C. sculpturatus* and *C. limpidus* are dangerous; their respective congeners, *C. pantherinus* and *C. vittatus* are not. There is no good correlation between size and potential danger to man among scorpions. Some very large species such as *Heterometrus cyaneus* of Java and *Hadrurus arizonensis* of the southwestern United States never cause serious envenomation, while small species such as *Centruroides sculpturatus* may be deadly.

Scorpion poisoning is an appreciable medical problem in some

Figure 7. Palestine Yellow Scorpion *(Leiurus quinquestriatus),* a dangerous species found in the Middle East and North Africa. Cardiac damage has been reported following its sting, and fatalities are not uncommon. (Photo by Illustration Department, Indiana University Medical Center.)

Figure 8. Lobster-clawed Scorpion *(Diplocentrus bigbendensis)* is native to the Big Bend region of Texas and adjacent Mexico. It has powerful claws, but its venom has little toxicity for man. (Photo by Illustration Department, Indiana University Medical Center.)

parts of the world, children being chiefly affected. Prior to 1958 more than a thousand deaths a year in Mexico were ascribed to this cause. The highest death rates are reported in the west central part of the country in the states of Colima, Nayarit, Guerreo, and Morelos (Mazotti and Bravo-Becherelle, 1963). In 1964 the state of Guanajuato reported 2394 scorpion stings with three deaths. *Centruroides suffusus* and *C. limpidus* are the most dangerous species in this region with *C. noxius* also of importance. Scorpion stings are not uncommon in the southern and western United States; however, nearly all serious and fatal ones occur in Arizona, most of the state being inhabited by the dangerous *Centruroides sculpturatus* which barely enters adjoining California and New Mexico. (*C. gertschi,* often listed as a second dangerous species of the region, appears to be merely a variety of *sculpturatus.*) According to a survey published in 1950, 1573 cases of scorpion sting were treated by Arizona physicians during a 10-month period, and there were 64 deaths between 1929 and 1948 (Stahnke, 1950). Since then, deaths have been less than one per year despite a more than four-fold increase in the state's population. The number of stings treated has also declined markedly. As judged by antiserum usage, there are no more than 40 to 60 serious cases annually (Kossuth, personal communication, 1972).

The most extensive region of high morbidity and mortality from scorpion sting extends from North Africa through Pakistan and the more arid parts of India. *Androctonus australis* and *Buthus occitanus* are the most dangerous species in the Saharan part of this region. In the Middle East, *Leiurus quinquestriatus* is the leading cause of fatalities, but dangerous species of *Androctonus* also occur. *Buthus tamulus* is the most dangerous species of peninsular India and occurs north to Sind. Detailed information on mortality and morbidity are available for only a few localities. In southern Algeria, 20,164 cases with 386 deaths were reported during a 17-year period (Balozet, 1964). Eighty-two patients, nine of whom died, were admitted to the Negev Central Hospital, Israel between 1960 and 1968 (Gueron and Yaron, 1970). In the Kobala District of Bombay State, Mundle (1961) saw 78 hospitalized cases in 14 years; there were 23 fatalities, nine of them adults. Serious and fatal scorpion stings are known from south India, Pakistan, and Iran, but figures on incidence are not available.

Over much of tropical South America and on a few West Indian islands, scorpions of the genus *Tityus* cause a significant number of severe cases of poisoning. In the city of Ribeirao Preto in southern Brazil, 1331 stings with eight deaths were reported during a seven-year period. Most were ascribed to *T. serrulatus,* the common domestic scorpion of that region (DaSilva, 1952). One hospital in Trinidad treated 698 cases with 33 fatalities between 1929 and 1933 (Waterman, 1938). Stings are still fairly frequent on that island, but the case fatality rate has been markedly reduced. In South Africa, scorpions of the genus *Parabuthus* occasionally cause severe stings and a few fatalities.

In Mexico, scorpion stings are most frequent from April through July, a period that includes the onset of the rainy season (Mazotti and Bravo-Becherelle, 1963). The peak in southern Brazil coincides with the onset of the summer rains in December. In India and Pakistan also, most stings seem to occur with the coming of the rains with a secondary rise at the end of the monsoon season. In northern Algeria and also in Arizona, the highest incidence of stings is May through August (Boisset and Larrouy, 1962; Stahnke, 1956). In Trinidad the most stings occur during June and July. Human dwellings, particularly those of adobe or rough stone construction, are attractive to scorpions including such dangerous species as *Tityus serrulatus, Buthus occitanus, Androctonus australis,* and *Centruroides suffusus.* Here they find shelter, moisture, and equitable temperature, and plentiful food in the form of roaches and other insects. In Arizona, *Centruroides sculpturatus* is most often found under bark of eucalyptus and cottonwood trees but occasionally invades modern, well constructed houses. Like some other scorpions, it tends to cling to the underside of objects in contact with the ground or floor (Stahnke, 1956). *Tityus trinitatus* is uncommon in houses but often abundant in cane fields and banana and cocoa plantations. Adult scorpions' ability to endure long periods without food or water enables them to be transported readily in timber and agricultural products and presents a risk to those handling such materials.

Stings of most species of scorpions cause only local pain and edema, sometimes with a little ecchymosis. These subside in a few minutes to a few hours even without treatment. If an individual is stung by one of the more dangerous species such as *Centruroides sculpturatus* or *Leiurus quinquestriatus,* the initial pain is followed by tingling or burning sensations radiating along the nerves from the site of the sting, but edema is rarely seen. Salivation, sweating, difficulty in speaking and swallowing, nausea, vomiting, hyperactivity, and exaggerated reflexes are characteristic; convulsions are frequent in small children. An initial brady-cardia is followed by tachycardia. Cardiac irregularities and ab-normal electrocardiograms sometimes simulating myocardial in-farction have been reported, particularly with *Leiurus* and *Tityus* stings. In Trinidad, 34 of 45 hospitalized patients showed signs of toxic myocarditis. All returned to normal in six days. Cardiac symptoms are believed to result from release of excess catechola-mines (Gueron and Yaron, 1970; Poon-King, 1963). Sting by *Tityus trinitatus* is a common cause of acute pancreatitis in Trini-dad. On the basis of elevated serum amylase, hyperglycemia, and other abnormal laboratory findings, it was diagnosed in about 75 per cent of patients hospitalized with scorpion stings. Only about half of these patients had abdominal pain or vomiting, and all recovered without evidence of permanent damage. The patho-genesis of the pancreatitis is not well understood (Bartholomew, 1970). Pulmonary edema often develops in serious scorpion en-venomation by any species and is a characteristic finding at au-topsy. Multiple internal hemorrhages and intravascular thrombi have been reported following fatalities due to *Buthus tamulus* (Devi et al., 1970). Deaths usually occur 4 to 72 hours after stings. Most deaths are in children under six years of age. *Parabuthus* stings cause marked stiffness of limbs and joints (Grasset et al., 1946).

Differential diagnosis in scorpion sting usually involves dis-tinguishing this type of injury from other venomous bites and stings as well as distinguishing stings by dangerous scorpions from those of the less dangerous species often found with them. The puncture made by a scorpion sting is almost invariably single, small, and bleeds little if at all; snakebites show larger, usually multiple punctures and bleed freely. Scorpion stings usually pro-duce more severe immediate pain than spider bites. Differenti-ation from insect stings may be impossible in early stages. Venom of *Centruroides sculpturatus* causes marked hyperesthesia. A use-ful diagnostic procedure is to tap the site of the sting gently. If the injury was caused by one of the more dangerous scorpions, there will be a prompt withdrawal reaction. This is not usually seen with insect stings or stings by less dangerous species of scorpions (Stahnke, 1972). Excessive salivation and sweating are also considered strongly suggestive of systemic scorpion poisoning.

Most scorpion stings require only symptomatic local treatment and reassurance. Prompt application of a ligature followed by an ice pack or immersion of the bitten part in ice water may retard absorption of venom and prevent or reduce the severity of a systemic reaction. Chilling should be maintained two to three hours; however, the ligature should be removed after about five minutes (Stahnke, 1956). Intravenous barbituates are the best drugs for controlling excitability and convulsions. Atropine and sympatholytics, e.g. phentolamine, are indicated in severe poisoning (Bursoum et al., 1954). Meperidine and morphine derivatives are contraindicated, and intravenous fluids should be given with caution because of the risk of pulmonary edema. Scorpion antivenins are available in most countries where dangerous scorpions occur. Ideally they should be specific for the genus involved, but animal experiments indicate that there is a considerable degree of cross-protection. Mexican anti-*Centruroides* serum neutralized North African *Buthus occitanus* and South American *Tityus* venoms effectively but not *Androctonus crassicauda* venom, while *A. crassicauda* antivenin neutralized *B. occitanus* and *Tityus* but not *Centruroides* (Whittemore et al., 1964). *Tityus* antivenin did not neutralize *Parabuthus* venom (Grasset et al., 1946).

Since scorpions can hide in inaccessible places for a long time, their eradication from around dwellings is more difficult than that of most household pests. The chlorinated hydrocarbon insecticides such as DDT, chlordane, lindane, and dieldrin are effective against scorpions; however, their overall ecological effect must be considered. A mixture of crankcase oil and creosote poured between the earth and house foundations will repel scorpions (Stahnke, 1956). Cats, solfugids, and certain lizards are among the natural enemies of scorpions, but their importance in control is difficult to assess.

II. SPIDERS

Spiders are amazingly successful and abundant animals. There are about 100,000 species, world wide in distribution and known from permanent snow fields atop the highest mountains to the lowest and hottest deserts. Aside from man, they are the only animals that have been able to travel by air without special anatomical modifications for flying. Their numbers may be prodigious; an estimate of the population per acre in a grassy field is 2,200,000. Since they are entirely carnivorous, an enormous number of insects and other small creatures is necessary to support such a population. One authority estimates British spiders eat each year a mass of prey equal to the weight of the human population of the British Isles (Bristowe, 1947). A great many spiders are arboreal, and many others are burrowers, but only a few have become to any degree aquatic. Numerous species adapt readily to living in close association with man. There are both diurnal and nocturnal species.

Male spiders are generally smaller than females and often differ in color and pattern. Among the widow spiders of the genus *Latrodectus,* males are too small to be venomous to man. This is not the case, however, with other medically important genera. In one *(Atrax)* males have a considerably more toxic venom and have accounted for most human fatalities. In all male spiders, the last palpal segment is enlarged forming a bulb-like structure used to inseminate the female. Nesting habits of spiders show many interesting adaptations. The female often gives a considerable degree of protection to the eggs and newly hatched young. Spiders have highly developed silk glands and make many uses of their secretion. Webs are the most universally familiar, but the silk is also used for egg sacs, guide lines, the lining of tunnels and burrows, and as a float for aerial travel.

All species of spiders have fangs and most have venom glands, but the vast majority are unable to injure man. Evidence is accumulating, however, that any spider with fangs capable of penetrating human skin can cause at least local envenomation. Some fifty species of spiders have been implicated in cases of human envenomation in the United States and numerous others in other parts of the world.

The venom apparatus of spiders is associated with the chelicerae or first pair of appendages of the head. Each consists of a large basal segment and a terminal claw-like fang pierced by the duct of the venom gland. In tarantulas and their kin (mygalomorphs), the fangs move parallel to each other in vertical planes, while in other spiders they move obliquely inward and backward toward each other. The saccular venom glands are lined with columnar secretory epithelium and are invested with a spiral layer of muscle and an outer sheath. In most spiders they lie in the anterior cephalothorax; in mygalomorphs they occupy the basal segments of the chelicerae. In the black widow, *Latrodectus mactans,* at least, there are two kinds of accessory glands. The secretion of one mixes with that of the principal venom gland and is stored in its lumen. The other type of accessory gland lies near the end of the venom duct, and its lipid-rich secretion mixes with that of

Figure 9. Fangs and left venom gland of Black Widow Spider *(Latrodectus mactans)*. Fangs of this type are characteristic of all spiders except the tarantulas and other mygalomorphs. (Photo by Illustration Department, Indiana University Medical Center.)

Figure 10. Head region of Brown Recluse Spider *(Loxosceles reclusa)* show-ing the three pairs of eyes characteristic of this genus and the comparatively small fangs. (Photo by Illustration Department, Indiana University Medical Center.)

the other glands at the moment of venom ejection. Venom ejection is under control of the spider, and species that ensnare their prey often do not use venom to kill it. Venom is usually ejected in defensive biting. Electron micrographs by Smith and Russell (1967) show the components of *Latrodectus* venom segregated as droplets within the cells. These droplets, enclosed in membranes, are secreted into the lumen of the gland. Histochemical studies of the venom glands of seven species of spiders by Arvy (1966) showed two types of secretory cells, one producing a granular material high in aromatic amino acids and the other nongranular product rich in sulfhydryl groups.

Venoms of larger spiders can be obtained by electrical stimulation of the venom glands and collected from the tips of the fangs. An alternative method is to dissect out and macerate the venom glands. The venoms are clear or faintly turbid liquids. Most are neutral or slightly alkaline; a few are definitely acid. Venom yields from some representative species of spiders are given in Table III.

Studies on the chemistry and pharmacology of spider venoms have been almost exclusively confined to a few genera of medical importance. Since spiders, like scorpions, feed almost entirely on the juices and liquified tissues of their prey, spider venoms may have originally been digestive secretions and still retain this function to some extent. Norment and Vinson (1969) found *Loxosceles reclusa* venom had a powerful lytic action on the muscle and fatty tissue of the tobacco budworm larva as well as on the hemocytes and hemolymph. A proteolytic enzyme highly active against casein and having a molecular weight of 10,840 was isolated from the venom of the South American tarantula *Pamphobeteus roseus* (Mebs, 1970).

Most of the active components of spider venoms are proteins. Fontali and Grasso (1964) separated *Latrodectus mactans tredecimguttatus* venom into three fractions, two of which were toxic for the housefly and one for the guinea pig. These fractions were heat labile and inactivated by mercaptoethanol but not by versene. McCrone and Hatala (1967) found the venom of *Latrodectus m. mactans* to be composed of seven protein and three nonprotein

Venom Diseases

TABLE III. YIELD AND TOXICITY OF SOME
MEDICALLY IMPORTANT SPIDER VENOMS

Species	Locality	Venom yield (mg)	Mouse LD_{50} $(mg/kg)^a$	Reference
Atrax robustus ♂	Australia	0.810	2.50 I-V	Wiener, 1959
A. robustus ♀		2.050	16.250 I-V	
Steatoda paykulliana	Yugoslavia	0.2-0.3 (wet venom)		Maretic et al., 1964
Latrodectus m. mactans	Florida (U.S.)	0.190	1.30 I-P	McCrone, 1964
L. m. tridecimuguttatus	Israel	0.238	0.59 I-P	″ ″
L. geometricus	Florida (U.S.)	0.097	0.43 I-P	″ ″
L. variolus	Florida (U.S.)	0.254	1.80 I-P	″ ″
L. bishopi	Florida (U.S.)	0.157	2.20 I-P	″ ″
L. hasseltii	Australia	0.120	1.00 I-V	Wiener, 1956
Phoneutria nigriventer	Brazil	1.160	0.34 I-V 0.70 S-C	Bücherl, 1953
Loxosceles laeta	Uruguay	0.350	16.0 S-C	Mackinnon and Whitkind, 1953
Loxosceles rufipes	Uruguay	0.375	40.0 S-C	″ ″
Lycosa erythrognatha	Brazil	1.000	62.50 S-C 0.40 I-V	Bücherl, 1964
Pterinochilus sp.	E. Africa	5.000	3.35 S-C	Freyvogel et al., 1968
Aphonopelma sp.	SW. U.S.		14.14 S-C	Stahnke and Johnson, 1967

^{a}I-V = intravenous.   S-C = subcutaneous.   I-P = intraperitoneal

fractions with all the toxicity for mammals limited to one protein
fraction. This fraction was slowly dialyzable, highly labile and
had a molecular weight of approximately 5000. Russell and Buess
(1970) studying geographic variation in this species by gel electrophoresis found 13 to 17 protein bands in their venom samples.*

*A most regrettable state of confusion exists regarding the taxonomy of the
well known and medically important *Latrodectus* spiders. McCrone and Levi (1964)
recognized four species, *L. mactans mactans*, *L. variolus*, *L. geometricus*, and *L.
bishopi* as native to the continental United States. Some recent workers have refer-
red populations of *mactans* from the western U. S. to the taxon *hesperus*, treating
it in some cases as a species and in others as a subspecies of *mactans*. The forms
tredecimguttatus (southern Europe and Middle East), *indistinctus* (South Africa)
and *hasseltii* (Australia and much of Asia) have variously been considered full species, subspecies of *L. mactans*, and variants of *mactans* not deserving nomenclatural
recognition. *L. curacaviensis* is a West Indian species closely allied to *variolus* and
bishopi. *L. pallidus* and *L. revivensis* are little known Middle Eastern species. The
taxonomic status of populations in many areas, e.g., northern Argentina, remains
unsettled. Fortunately for clinicians, *Latrodectus* bites in man produce much the
same symptoms and respond to the same treatment.

Venoms of *Loxosceles* spiders contain four to six protein components, some of which appear to be responsible for their toxicity for mammals (Smith and Micks, 1968; Suarez et al., 1971). A protein toxin immunologically identical with that of *Centruroides sculpturatus* was reported in the venom of a tarantula of the genus *Aphonopelma* (Stahnke and Johnson, 1967). Venoms of the South American spiders *Phoneutra fera* and *Lycosa erythrognatha* contain toxic basic polypeptides with molecular weights of 5000 to 6000 and are high in histamine and certain free amino acids, especially glutamic acid (Fischer and Bohn, 1957; Schenberg and Lima, 1966). Venom of the large African spider *Pterinochilus* was separated into 24 fractions of which four proteins contained most of the toxicity (Freyvogel et al., 1968). Venom of the Australian mygalomorph, *Atrax robustus* has been fractionated into several components including non-toxic proteins. About half the venom consists of compounds with molecular weights of less than 500. These cause smooth muscle contraction and account for some of the toxicity of the venom. Most of the toxicity, however, resides in a fraction with a molecular weight of 15,000 to 25,000 as estimated by membrane filtration. It causes repetitive contraction of skeletal muscle and duplicates most of the symptoms produced by crude venom (Wiener, 1963; Sutherland, 1972). Enzymes including proteases, phosphodiesterase, and hyaluronidase have been found in several spider venoms (Kaiser and Rabb, 1967; Russell, 1966; Kaiser, 1956). Serotonin was found in the venom, venom glands, or chelicerae of six of eight species of spiders and was particularly high in *Phoneutria fera* (Welsh and Batty, 1963).

Latrodectus venoms are highly toxic for mammals, the horse and camel being particularly susceptible and the dog much less so. In the rat and guinea pig, envenomation is characterized by local irritation and muscle spasm, ataxia, tonic and clonic convulsions, excessive salivation, and flaccid paralysis. The blood pressure is increased and abnormalities in the electrocardiogram similar to those produced by digitalis and barium are seen. When spiders whose venom was labelled with ^{32}P were allowed to bite guinea pigs, the highest concentrations of the isotope were in the central nervous system, the peripheral nerves, and the muscles around the site of injection. Small amounts were found in the blood, lungs, heart, liver, and spleen (Maretic, 1963; McCrone and Porter, 1965; Lebez et al., 1965).

Cold-blooded vertebrates generally have a high resistance to *Latrodectus* venoms, but I have seen a frog die as a result of a bite by *L. mactans*. On a weight-for-weight basis, the housefly is about as susceptible as the mouse; however, the roach, *Periplaneta*, is about 30 times as resistant.

Longenecker et al. (1970) found *Latrodectus* venom applied to the frog neuromuscular junction caused exhaustion of miniature end-plate activity and depleted the nerve terminals of vesicles. Washing did not reverse the activity, but *Latrodectus* antivenin did. The nerve terminal membrane was believed to be the site of action of the venom. In preparations from the roach nerve cord, the venom blocked synaptic transmission between cercal nerves and giant fibers without impairing axonal conduction. The venom may act by causing a massive release of acetylcholine and a subsequent depolarization of the postsynaptic membrane (D'Ajello et al., 1969, 1971).

Venoms of the recluse spiders, *Loxosceles*, are best known for their powerful dermonecrotic and hemolytic activity in mammals, but their mode of action is little known. Among the common laboratory animals, rabbits and guinea pigs are highly susceptible, rats quite resistant. Experimentally-induced spider bites or intradermal injection of venom in sublethal doses causes production of a wheal followed by a black, necrotic eschar that eventually sloughs and heals. There is a strong tendency for the lesion to spread downward as though by gravity. Histologically, dilated blood vessels and hemorrhage are seen early; necrosis, leukocytic infiltration, and thrombosis of capillaries with hyaline material later. The necrosis may extend into the underlying muscle. If larger doses are given, the animal dies with profuse hemorrhages around the site of inoculation and in the skeletal muscles, gastrointestinal mucosa, and thymus (Atkins et al., 1958). The lesion produced by intradermal injection of venom resembles the Arthus reaction. Rabbits depleted of neutrophils responded with much less necrosis and edema; animals depleted of complement with less hemorrhage and edema. Apparently both neutrophils and complement are necessary for full development of the dermal lesion (Smith and Micks, 1970). Intravenous injection of venom into dogs is followed by intravascular hemolysis and thrombocytopenia. If sublethal doses are given, the hematologic picture returns to normal in three to five days. *In vitro* the venom lyses human erythrocytes. The hemolysin is not dependent on complement and is inactivated by heat (Denny et al., 1964). The venom contains a potent inhibitor of the C'-5 component of human complement (Kniker et al., 1969).

Venom of the Brazilian wandering spider *Phoneutria fera* causes salivation and penile erection in mice in sublethal doses; larger doses cause dyspnea and paralysis of the hindquarters. Similar effects are seen in dogs; in addition there may be sneezing, lacrimation, mydriasis, vomiting, and blood in the feces. Histamine-like effects such as precipitous hypotension and contraction of the isolated guinea pig ileum are also seen. One polypeptide fraction produced flaccid paralysis of a type different from that produced by the whole venom. Most of the other activities were seen with more than one of the polypeptide fractions. Rats and rabbits are highly resistant to this venom (Schenberg and Lima, 1966).

Venoms of the large tropical tarantulas as examplified by *Pterinochilus* of Tanzania and *Hapalopus* of Peru produce little or no local reaction in guinea pigs, rabbits, dogs, and mice. An initial period of irritability and excitement is followed by weakness, dyspnea, and paralysis. Salivation, lacrimation, and convulsions were prominent with *Pterinochilus* venom. With both species, degenerative lesions were found in the renal tubules and liver; lesions in the brain, adrenals and myocardium were seen with *Pterinochilus;* and pulmonary congestion was prominent in animals receiving lethal doses of *Hapalopus* venom (Maretic, 1967; Espiñoza, 1966). Venom from tarantulas of the southwestern U.S. *(Aphonopelma)* produce generally similar effects on rats, although excitement seems to dominate the picture to a greater degree (Stahnke and Johnson, 1967). The venom of *Atrax robustus* also shows this general type of pharmacological activity;

in addition a marked drop in body temperature was noted and autopsy findings were confined to emphysema and pulmonary congestion often with hemorrhages. Mice, guinea pigs, and rabbits were relatively very resistant to this venom; however, a monkey succumbed promptly to a naturally-inflicted bite (Wiener, 1957).

Latrodectus envenomation in man is a medical problem in many parts of the world but chiefly in temperate and subtropical regions. In the dry coastal plains and low hills of the Mediterranean area and in parts of Russia, *Latrodectus mactans tredecimguttatus* characteristically inhabits grain fields. In years when the spiders are exceptionally common, veritable epidemics of arachnidism may occur during harvest time. Between 1946 and 1950 there were 492 reported cases in the Tyrrhenian coast of central Italy (Bettini, 1964). Numerous cases have also been reported from the Istria peninsula of Yugoslavia (Maretic, 1966). In South Africa bites by *L. m. indistinctus* are likewise usually associated with the wheat harvest, while bites by *L. geometricus* are more frequently sustained in vineyards or near buildings (Finlayson, 1955). The *Latrodectus* spiders of the United States and Mexico are most plentiful in crevices or under cover close to the ground; however, *L. variolus* is distinctly arboreal and *L. bishopi* somewhat so. Their webs are irregular and untidy, although sometimes quite extensive. The common black widow *(L. m. mactans)* is found over most of the continental U. S. and may be very plentiful in stone walls, trash piles, barns, stables, and outdoor toilets. It does not adapt well to living in occupied dwellings but may be brought indoors on garden produce, firewood, and similar materials. Although bites are more frequent in the southern states, there are no particular foci of concentration. Most bites occur in rural or suburban areas but in or close to buildings. Adult males are bitten somewhat more frequently than other age and sex groups.

Bites of the recluse spiders *(Loxosceles)* may well be of greater medical importance in the U. S. than those of black widows. The brown recluse *(L. reclusa)* has been reported from nearly half the states, although it is most plentiful in the mid-South and southern Midwest. Other species such as *L. unicolor* and *L. arizonica* inhabit the Southwest, and *L. laeta* and *L. rufescens* have been introduced and occur sporadically. The original habitat of these spiders was evidently under stones and bark in a relatively dry environment, but *laeta, reclusa,* and *rufescens* all adapt readily to living in houses and other buildings. They are secretive and often hide under boxes, in closets, and behind pictures and clothing hanging against the wall. Bites are more likely to occur indoors than out, and in many cases the spider bites when trapped under clothing or bedding. Children are bitten more frequently than adults and women more frequently than men (Gorham, 1968). Loxoscelism was first reported from Chile (Macchiavello, 1937) and continues to be reported frequently from that country as well

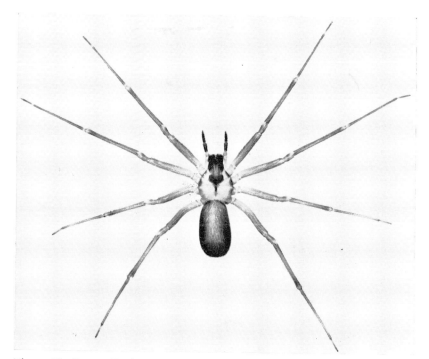

Figure 11. Brown Recluse Spider *(Loxosceles reclusa),* body length about 12 mm. This is the species most frequently implicated in cases of necrotic arachnidism in the United States. The dark mark on the cephalothorax characterizes most North American species of this genus. (Photo by Illinois Natural History Survey.)

Venom Diseases

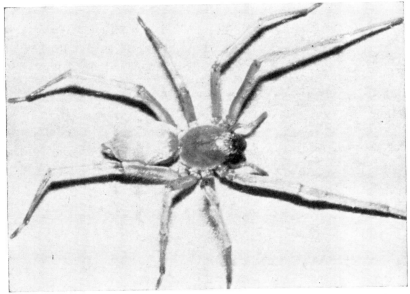

Figure 12. *Chiracanthium mildei,* a small pallid spider, body length about 7 mm. It is commonly found in buildings. This species and some others of its genus may cause moderately severe envenomation. (Photo by Illustration Department, Indiana University Medical Center.)

as from Peru, Bolivia, Argentina and Uruguay (Gajardo-Tobar, 1966). *L. laeta* is the species involved in most areas, but others such as *L. rufipes* occur. This species frequents dry, rocky habitats along the Peruvian coast but it is not uncommon in houses in Lima (Delgado, 1966). In the Old World, *Loxosceles* bites have been reported from southern Europe and the Middle East (Efrati, 1969).

The small, pallid spiders of the genus *Chiracanthium* have been incriminated in mild to moderately severe cases of arachnidism in southern Europe, Hawaii, and several of the mainland states. These spiders occur in both urban and rural areas and frequently inhabit houses and other buildings. The epidemiological pattern of bites is much like that for *Loxosceles.*

Comparatively little is on record concerning the epidemiology of bites by the tropical American wandering spiders of the genus

Phoneutria. Most bites have been reported from Brazil and northern Argentina, apparently in a rural environment. The tropical wolf spiders, tarantulas, and the Australian funnel-web spiders are outdoor species and show no particular preference for the vicinity of human habitations, however *Atrax robustus* may be found in the suburbs of Sydney, and the large tarantulas are often found in banana plantations and other cultivated areas.

Many kinds of spiders are readily transported by man. Two species of *Loxosceles* and two of *Chiracanthium* are among the medically important spiders that have been introduced into the United States and become established, at least locally. *Loxosceles reclusa* has been recorded from seven localities beyond its presumed natural range (Gorham, 1968). The part played by man in the dissemination of the various forms of *Latrodectus* is unknown; however, the American black widow has recently been reported from Israel (Shulov, 1966). A large and mildly venomous crab spider, *Heteropoda venatoria,* is abundant in many tropical and subtropical seaports including some in the southern United States and is sometimes transported to northern cities. It is the "leopard tarantula" of fruit dealers. True tarantulas were formerly found with regularity in banana shipments.

Few spiders are to any degree aggressive toward animals too large to serve them as food, and it is very difficult to induce many species to bite under experimental conditions. Most bites inflicted on man occur when the spider is accidentally or intentionally injured or restrained. Black widows, however, sometimes attack when their webs are disturbed. Wandering spiders *(Phoneutria)* and funnel-web spiders *(Atrax)* have also been reported to react aggressively toward human molestation.

Spider bites do not always result in envenomation. The fangs may be too weak to penetrate the skin, or the arachnid may fail to inject venom. This is frequently the case with some of the large tarantulas. In a series of animal experiments with *Pterinochilus,* 24 of 33 bites failed to produce detectable toximia (Maretic, 1967).

Within the genera of medically important spiders, the manifestations of envenomation in man seem to be qualitatively much the same; the differences reported apparently reflect the amount of venom injected rather than interspecific differencs in type of venom. With *Latrodectus* spiders, the bite causes mild to moderate pain that usually disappears promptly and no other local reaction. Symptoms of general toxemia begin after a short latent period. The most characteristic of these are muscle spasm and cramps that may involve any part of the body but frequently result in marked abdominal rigidity without tenderness. Pain is intense and often accompanied by a high degree of anxiety. Paraesthesia, frequently described as a burning sensation, may affect the entire body but is often most severe on the soles of the feet. Headache, sweating, nausea, and facial congestion are frequent. Edema of the eyelids and scarletiniform rash are occasionally seen. The blood pressure is usually normal or elevated, but profound shock has been reported. Abnormalities in the electrocardiogram have been observed. The temperature is slightly to moderately elevated, tendon reflexes are hyperactive, spinal fluid pressure increased and urinary output decreased. Leukocytosis and albuminuria are common laboratory findings. Complete recovery after a few days illness is the rule. Deaths are infrequent and usually occur among the very young, the aged, or those with hypertension (Frank, 1942; Greer, 1949; Maretic, 1966).

This type of arachnidism may easily be confused with acute appendicitis, perforated ulcer, or other abdominal emergency particularly if a clear-cut history of spider-bite cannot be obtained. The cutaneous hyperaesthesia and paraesthesia and increased tendon reflexes are useful diagnostic signs pointing to spider envenomation.

Phoneutria spiders' bites are very painful perhaps because of the high serotonin content of the venom. Cutaneous hyperaesthesia and muscle spasms are a prominent part of the general picture, but convulsions are more frequent than in *Latrodectus* envenomation and may progress to opisthotonus in severe cases. The temperature is subnormal, the pulse rapid and irregular, and oliguria or anuria is often seen. Most deaths are in children and occur during the first five hours of illness. Most patients recover in six to 12 hours, but a zone of paraesthesia around the bite may persist several days (Gajardo-Tobar, 1966).

Loxosceles bite produces a distinctly different clinical syndrome from that of *Latrodectus*. Some patients report a fairly severe stinging sensation when bitten, for others the bite is not painful enough to be noticed at the time. A bluish-white halo of vasoconstriction surrounds the bite during the first few hours and becomes a blister surrounded by a painful, reddish zone with irregular margins. This may be accompanied by chills, malaise, and a generalized rash. The hemorrhagic zone may continue to spread over a period of three to six days, and most of it becomes necrotic with destruction of the skin and subcutaneous fat. Healing is slow and results in an extensive scar. Anderson (1971) reports the tissue pathology resembles that of the Shwartzman reaction and may be due to local activation of clotting mechanisms with capillary thromboses and ischemia. Fibrinogen, thrombin, and platelet values remain normal, however. In a few cases, usually in children, massive intravascular hemolysis develops soon after the bite accompanied by hemoglobinuria, jaundice, high fever, and shock. Such cases have a grave prognosis; Latin American physicians report a mortality of about 30 per cent with renal shutdown or pulmonary edema as common terminal events. Several fatalities have occurred in the United States (Schenone and Prats, 1961; Dillaha et al., 1964; Taylor and Denny, 1966). The pathogenesis of the hemolytic reaction is not understood. The comparatively weak hemolysin demonstrable *in vitro* scarcely seems adequate by itself to account for the massive destruction of erythrocytes observed clinically.

A fatal case with both hemolytic and hemorrhagic manifestations was recently reported. Platelets and fibrinogen were decreased, petechiae were observed, and hemorrhages in the kidney and subdural space were seen at autopsy. Hemoglobin was present in quantity in the lungs, kidneys and spleen. Death occurred about 30 hours after the bite despite administration of corticosteroids, heparin, and whole blood (Vorse et al., 1972).

Severe loxoscelism has so far been associated only with bites of *L. laeta* and *L. reclusa*. Cases caused by *L. unicolor* and *L. arizonica* in the southwestern U. S. (Russell et al., 1969) and *L. rufescens* in Israel (Efrati, 1969) have generally been milder.

Necrotic arachnidism in South America has also been attributed to bites of the wolf spider, *Lycosa raptoria,* and the catheaded spider, *Glyptocranium gasteracanthoides* (Gajardo-Tobar, 1966).

Bites of *Chiracanthium punctorium* and *C. inclusum* have been reported as causing varying degrees of local pain, redness, edema, and lymphadenopathy occasionally accompanied by chills and low fever (Maretic, 1962; Furman and Reeves, 1957). *C. mildei* bites may resemble mild *Loxosceles* envenomation (Spielman and Levi, 1970; Minton, 1972).

Bites by the large wolf spiders of the genus *Lycosa* are occasionally reported in the United States, Europe, and South America. In most instances they cause only local pain, swelling, and ischemia of brief duration. Local necrotic and systemic reactions have been reported, however (Grothaus and Teller, 1968; Maretic and Lebez, 1970). Bites of the golden orb-weaver *(Argiope aurantia)* a common large spider of the eastern United States, and *Phidippus formosus,* a western jumping spider, have been associated with mild envenomation and largely local symptoms (Gorham and Rheney, 1968; Russell, 1970). A unique type of injury due to ejection of venom into the eye was caused by the green lynx spider *(Peucetia viridans)* (Tinkham, 1946).

Necrotizing spider bites must be differentiated from streptococcal cellulitis and other types of bacterial infection including anthrax and rickettsial diseases in some parts of the world. The patient may not recall the bite or may attribute the lesion to some other cause. The characteristic evolution of the lesion plus negative bacteriological findings, at least in the early stages, usually permit diagnosis. Other types of arthropod bites must also be differentiated from arachnidism. Bedbug, flea, and mite bites are usually multiple; spider bites rarely so. Bites of venomous *Ornithodorus* ticks and triatomid and reduviid bugs may be very difficult to differentiate from spider bites. If inflicted indoors, a careful search of the premises may uncover the culprit. The full-blown picture of loxoscelism or latrodectism does not develop, but no characteristic features accompany bites of the less venomous spiders.

Bites of Australian funnel-web spiders *(Atrax robustus* and

Figure 13. Golden Garden Spider *(Argiope aurantia)*, a common orb-weaver of the eastern U. S. Females have a body length of about 25 mm. The zig-zag silk reenforcement of the web is characteristic. Bites by this spider have been reported but do not cause serious injury. (Photo by author.)

Figure 14. Unidentified tarantula photographed in eastern Mexico; body length about 60 mm. These large tropical spiders rarely bite man, and there are few reports of serious envenomation. (Photo by author.)

A. formidabilis) cause sharp local pain quickly followed by weakness, sweating, and nausea. The temperature is elevated and pulse and respirations increased. The pupils are small and fixed and tendon reflexes diminished. Pulmonary edema accompanied by cyanosis and dyspnea may develop. Ten fatalities, half of them in children under 10, were recorded from this type of envenomation between 1927 and 1960. All but one of the deaths took place within 12 hours after the bite; in other cases recovery was prompt and complete (Ingram and Musgrave, 1933; Wiener, 1961). Sutherland (1972) reported the death of a young woman and her unborn child from the bite of this spider in 1971 and mentioned other near fatal cases.

Information concerning bites by the large tarantulas of tropical America, Africa, and Asia is meagre. This is probably in large part due to the nocturnal habits, comparative rarity, and inoffensive disposition of these spiders. Cases of bites by *Pternochilus* in Tanzania (Freyvogel et al., 1968) and by *Lampropelma* in Malaya (Lim and Davie, 1970) resulted in brief, comparatively mild envenomation with mostly local manifestations. Nevertheless, the venoms of some of these spiders have relatively high toxicity for animals and the creatures are feared by natives of regions where they occur, so it is possible that some bites are serious. The common tarantulas of the southwestern United States *(Aphonopelma)* have weak venom and are exceptionally inoffensive. No authentic case of severe envenomation from their bite is on record.

Treatment of spider bite makes use of both immunotherapy and pharmacological antagonists to the effects of the venom. Early first aid measures are of questionable value, although local application of cold as described under treatment of scorpion sting may be of some help. Surgical excision of the bite site has been suggested for *Loxosceles* bites seen and definitely diagnosed during the first two hours.

Intravenous calcium gluconate (10 ml of a 10 per cent solution) is usually effective in relieving the muscle cramps, headache, nausea and most of the other distressing symptoms of *Latrodectus* envenomation, but its action is brief, and subsequent doses often have less effect than the original. Better results are achieved by use of muscle relaxants such as methocarbamol given in an initial intravenous dose of 10-20 ml followed by administration by mouth (Russell, 1962). Persons bitten by *Loxosceles* spiders should receive a moderately large dose of corticosteroid (e.g. 100 mg prednisone or 16 mg dexamethasone) as soon as there is definite evidence of envenomation and daily for the first few days tapering off gradually depending on the response. Corticosteroids have been effective when begun as late as 72 hours after the bite. They have not always been successful in preventing either hemolytic crisis or extensive necrosis, but this may have been because of insufficient dosage. Maintenance of an alkaline urine and transfusions are beneficial in patients with intravascular hemolysis.

Heparin produced marked improvement in one severe case reported by Anderson (1971).

Although immunological differences among venoms of *Latrodectus* spiders were found by McCrone and Netzloff (1965), these appear to be of little clinical significance. The important toxin is apparently the same in all species. *Latrodectus* antivenins are produced in the United States and in several other countries. In U. S. practice, antivenin is generally reserved for children or patients with cardiovascular disease. A single intravenous or intramuscular dose of 2.5 ml is generally sufficient. *Phoneutria* antivenin is produced by the Instituto Butantan in Brazil and is the treatment of choice for bites by the dangerous spiders of this genus. *Loxosceles* antivenin is also produced by this institute but has not received adequate clinical evaluation. Smith and Micks (1968) reported immunological differences among the venoms of *L. reclusa, L. laeta,* and *L. rufescens,* those of *reclusa* and *rufescens* being closer to each other than to *laeta.* The clinical implications of these observations are unknown, but the similarity of manifestations in *Loxosceles* poisoning by any species suggests that the toxins may be much the same.

Since the toxins of *Atrax* venom are nonantigenic and no adequate pharmacological antagonist to their effects has been discovered, treatment of bites by these spiders is symptomatic. Since the effects of the venom are short-lived, use of a respirator and other supportive measures should be employed in severe envenomation.

Control measures against spiders are usually practical only in the eradication of *Loxosceles* and *Latrodectus* from dwellings and other occupied buildings. Lindane, chlordane, and malathion are all effective but should be used with caution. Spiders in small, enclosed spaces such as closets and boxes may be killed by paradichlorbenzene or dichlorlorvos-impregnated plastic strips. Destruction of roaches, silverfish, and other household insect pests deprives spiders of their food supply. Frequent vacuum cleaning of those parts of the house likely to harbor spiders and elimination of boards, trash, and other ground cover from around houses may be helpful. Natural enemies of adult spiders include sphecid and pompilid wasps, ants, some birds, lizards, and small snakes. The small wasps, *Gelis niger* and *Pimplaoculatoria,* heavily parasitize the egg cocoons of *Latrodectus m. tredecimguttatus* in southern Europe (Maretic, 1966). Similar parasitism in the United States has been observed with the scelionid wasp, *Baeus latrodecti,* and *Gaurax aranease,* a small dipteran.

III. MISCELLANEOUS GROUPS

1. Centipedes

Centipedes are elongate, many-segmented arthropods with one pair of legs to each typical body segment. They are virtually worldwide in distribution, although all the large species occur in the tropics and warm temperate regions. Most are secretive and nocturnal. Many climb well, and others are burrowers. With rare exceptions, they are predators, and the large species do not hesitate to feed on small vertebrates. Female centipedes of some

Figure 15. *Scolopendra* sp., specimen about 100 cm long collected in Trigo County, Kansas. Centipedes of this genus are widespread in tropical and warm temperate regions and are responsible for numerous cases of human envenomation. (Photo by author.)

Figure 16. Anterior end of large *Scolopendra* centipede (length about 18 cm) showing fangs. (Photo by author.)

families curl round their egg clusters and may actively defend them.

In centipedes the first pair of legs is modified to form two heavy, curved, hollow fangs. The cylindrical or pyriform venom glands occupy the bases of the fangs. The acini of the gland are disposed radially around a central canal that extends to the tip of the fang. Variations in morphology of the gland are seen especially in the burrowing forms such as *Geophilus* and *Chaetechelyne* (Phisalix, 1922, vol. 1, pp. 191-193). The venom is a clear, colorless liquid. There is virtually no information on venom yield of the various species.

Centipedes employ their venom to kill prey and for defense. It is also probable that the venom has a digestive function as well. The chemistry of centipede venom is unknown except for the observation that *Scolopendra viridicornis* venom contained a moderate amount of serotonin (Welsh and Batty, 1963).

Bücherl (1946) found the mouse LD_{50} of *Scolopendra viridi-*

cornis venom to be 0.03 gland intravenously and 0.25 gland sub-cutaneously. Venom of *S. subspinipes* was less toxic, the intravenous LD_{50} being 0.05 gland and the subcutaneous 0.7 gland. For *Otostigmus scabricauda,* the doses were 0.012 gland and 0.37 gland respectively. The largest species, *S. viridicornis,* killed mice by naturally inflicted bites in about three minutes and guinea pigs in four to twenty-four hours. Bites of *S. subspinipes* were lethal to mice but not guinea pigs, while bites of *Otostigmus* and two other species failed to kill mice. The animals appeared to die from respiratory paralysis, and rigor developed very quickly after death. Sublethal doses of venom caused local erythema and superficial necrosis.

Human envenomation by centipedes has been reported from the Philippines, Malaya, New Guinea, Crimea, Brazil, the south-western United States, and Hawaii. Under some circumstances it may be fairly common. A U. S. army unit on Leyte in the Philippines had many men bitten by large *Scolopendra subspinipes* that entered tents and latrines in search of insects (Remington, 1950). Almost without exception the bites are reported as being very painful and frequently followed by lymphangitis and lymphadenitis. Swelling and tenderness may persist up to three weeks after the bite. Necrosis occasionally occurs at the site of fang puncture. Haneveld (1956) stated that in 7 of 12 cases he treated in Dutch New Guinea (now West Irian), the pain and swelling subsided after treatment but returned about a week later and persisted one to three days. In a bite inflicted by the common North American house centipede *(Scutigera coleoptrata)* on a volunteer subject, the wheal and burning sensation disappeared within a few hours but returned about 24 hours later and persisted for about a day (Ewing, 1928). Haneveld attributed these recurrences to infection and bacterial allergy; however, confirmation by appropriate microbiological studies would be desirable. He reported paralyses, contractures, and heart irregularities in rare cases. Micks (1960) reported "severe systemic reactions" in two of eight cases from Texas; no details were given. There is a report from the Philippines of a child that died 29 hours after being bitten on the head (Pineda, cited by Remington, 1950). A fatal case in Arizona was ascribed to secondary infection (Stahnke, 1956).

While doing field work on Ponape in the Caroline Islands I was bitten twice near the tip of the forefinger by a *Scolopendra* 105 mm in body length. There was immediate pain and bleeding from the fang punctures. Forty-five minutes after the bite the entire finger was erythematous and almost too stiff to bend; the tip was very tense and somewhat painful. Four hours after the bite, function of the finger was normal, but the two distal phalanges remained swollen and sore. Pain and edema subsided over the next 20 hours leaving a tiny black eschar at the site of one fang puncture. No treatment was administered at any time.

The pain of centipede bites is readily relieved by infiltrating the bitten area with procaine or other local anaesthetic. Other treatment is rarely necessary, although Ariff (1956) advises the administration of 75 mg of cortisone or its equivalent during the first 24 hours after the bite. Infection may be somewhat more likely than in other arthropod envenomations, but routine use of antibiotics is probably unnecessary.

2. Millipedes

Millipedes differ from centipedes in having two pairs of legs to each typical body segment and in lacking structures for venom injection. Most millipedes are elongate, but some are short, wide, and capable of rolling up into a ball. They occur throughout most of the world but reach a greater size in the tropics and subtropics. Most species are secretive or fossorial. Their principal food is decaying vegetation; a few are scavengers or feed on living plants.

Millipedes do not bite, but many species have toxic and repugnatorial secretions that exude onto the body surface when the animal is disturbed. Generally a pair of glands is located in each segment, and only a few segments discharge their contents at a time. Benzoquinones and phenolic derivatives have been identified in these secretions. *Apheloria corrugata* produces hydrogen cyanide from precursors that are mixed just before discharge (Eisner and Meinwald, 1966).

Human injury from millipedes is infrequent. The large tropical species *Rhinocridium lethifer* can squirt an irritant fluid 25 to

50 cm. Droplets striking the eye produce violent conjunctivitis, and natives of New Guinea say that blindness sometimes results (Haneveld, 1958). An entomologist who put a large African milipede in his pocket for about an hour found that a large patch of skin on his thigh had turned black. It subsequently sloughed leaving a raw area (Burtt, 1947). A case of vesicular dermatitis in Nuevo Leon, Mexico was caused by secretions of a millipede of the genus *Orthopterus* (Halstead and Ryckman, 1949). Millipedes sometimes aggregate in enormous numbers, and the secretions of their repugnatorial glands can cause irritation of the mucous membranes and nausea.

3. Ticks

Ticks are a group of comparatively large Acarina which are ectoparasites on terrestrial vertebrates. In general, the Ixodid or hard ticks remain attached to the body of their host throughout most of their life cycle, whereas the Argasid or soft ticks live in the host's nest or den. Accordingly, the hard ticks feed on blood continuously; the soft ticks intermittently. The latter have amazing resistance to starvation and drying; engorged adults have lived months in vials without food or water. As a group, ticks are cosmopolitan. Species distribution depends in part upon the distribution of suitable hosts.

The biting and blood sucking apparatus of ticks consists of a prominent toothed hypostome and two hooked chelicerae which are inserted into the tissues. Salivary secretions are injected as the blood is withdrawn. Ticks are vectors of many diseases of medical or veterinary importance; in addition a true envenomation may accompany their bites as a result of injected salivary secretions. The venomous property of these secretions is evidently a biological accident, affording the tick no advantage either in defense or food getting.

The chemistry of tick saliva is essentially unknown. A paralytic toxin obtained from *Ixodes holocyclus* was nondialysable, inactivated within 15 minutes at 100°, and not affected by pepsin, papain, or trypsin. It could readily be separated by chromatography from the anticoagulant fraction (Kaire, 1966). The mode of action of the paralytic toxin is uncertain, although there is agreement from both experimental and clinical observations that there is a slowing of motor nerve conduction and reduction in amplitude of muscle action potential. In dogs there was inhibition of acetylcholine release at the neuromuscular junction, but no evidence of a defect in neuromuscular transmission was observed in a human patient (Murnaghan, 1960; Cherington and Snyder, 1968).

Tick paralysis is a unique form of envenomation affecting man and certain other mammals, notably cattle, horses, sheep, dogs, deer, and bison. Human tick paralysis has most frequently been

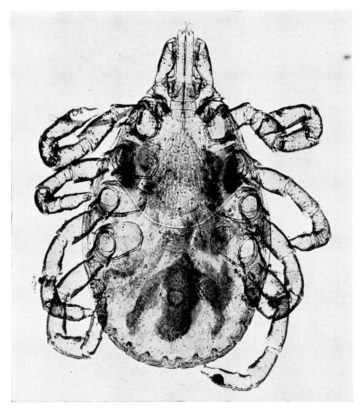

Figure 17. *Dermacentor* sp., ticks of this genus are the most frequent cause of tick paralysis in the United States. They are common parasites of wild and domestic mammals and frequently attach to man. ×20. (Photo by Illustration Department, Indiana University Medical Center.)

reported from the northwestern United States and adjacent Canada where it is caused by *Dermacentor andersoni* (Rose, 1954). Cases in the southeastern United States have been ascribed to *D. variabilis, Amblyomma americanum,* and *A. maculatum.* The condition has also been reported from Australia *(Ixodes holocyclus),* Crete *(I. ricinus* and *Haemaphysalis punctata),* and South Africa *(Rhinicephalus simus* and *Hyalomma truncata).* Most of the ticks found on patients with paralysis have been adult females, but cases caused by male ticks are on record (Campbell, 1964; Henderson, 1961). The toxin seems to be secreted in the ticks' saliva at a variable and irregular rate.

Although tick paralysis has been reported in adults of both sexes, about 65 per cent of cases are in girls under ten years old. This may merely mean that ticks in girls' hair are more apt to be overlooked; however, there may be a sex and age difference in susceptibility to the toxin. Persons living in or visiting rural areas are usually affected, and the incidence is highest in early summer.

There is reason to believe that ticks must be attached for four to five days before symptoms develop. Anorexia, irritability, and lethargy are noticed first and are followed by weakness, incoordination, and ataxia. Flaccid paralysis involves the legs, trunk, arms, and neck in that order. The reflexes are greatly decreased or absent. Sensory changes are minimal or lacking. There is little or no fever, and the spinal fluid is normal (Webb and Earnest, 1963; Sanzenbacher and Conrad, 1968). Fatal bulbar or respiratory paralysis may develop. Mortality in a series of 332 cases was 12 per cent (Rose, 1954). Removal of the tick, which is usually attached to the scalp or neck, is curative. Improvement may begin within an hour, and recovery is usually complete within 48 hours. A respirator and other supportive measures may be required in severe cases. The absence of fever, negative spinal fluid findings, and lack of any characteristic sensory involvement differentiate tick paralysis from infections of the central nervous system, myelitis, polyneuritis, and similar conditions.

Argasid ticks of the genus *Ornithodorus* are painful biters whose saliva is toxic. The "pajaroello" *(O. coriaceus)* of Mexico and the southwestern U. S. inflicts a bite that is followed by local

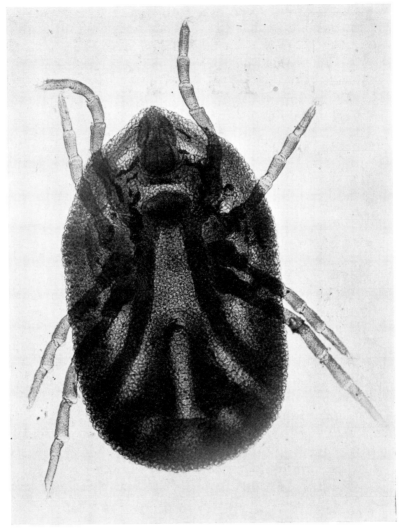

Figure 18. *Ornithodorus* sp., these ticks can inflict painful bites sometimes accompanied by local necrosis. The nature of the toxin in their saliva is unknown. ×20. (Photo by Illustration Department, Indiana University Medical Center.)

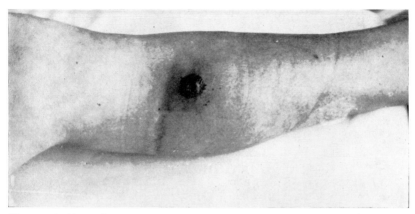

Figure 19. Necrotic lesion resulting from bite of the tick, *Ixodes ricinus*. This lesion is the result of envenomation and should not be confused with the eschar seen in some tick-borne rickettsial diseases. (Photo courtesy Z. Maretic.)

ecchymosis and ulceration. Swelling may involve an entire extremity and be accompanied by general malaise. Similar symptoms may follow bites by species of the *O. moubata* complex which frequently infest native dwellings in Africa and the Middle East. Larvae and nymphs of *O. stageri* which frequents bat caves in Texas and northern Mexico bite viciously leaving a welt that persists several days. The cosmopolitan ticks of the genus *Ixodes* frequently attack man inflicting painful bites sometimes accompanied by ulceration, fever, and vomiting (Eads et al., 1956; James and Harwood, 1969). There is no specific treatment for this type of tick envenomation.

Repellants such as indalone, dimethyl phthalate, and benzyl benzoate applied to clothing provide effective protection against ticks that may be encountered in the course of work or recreation. Where ticks are plentiful, children should be inspected for them daily. Ticks found attached should be painted with mineral oil, chloroform, or nail polish or touched with a hot object and then gently removed without crushing or tearing them. Treating vegetation with insecticides to control ticks is too expensive for widespread application and ecologically undesirable. Malathion can be used to treat infested premises.

Chapter 3

INSECT VENOMS AND ENVENOMATION

I. INTRODUCTION

IN NUMBERS OF SPECIES, insects surpass all other forms of multi-cellular life, and they may well exceed all other land animals in biomass. They occur in virtually all terrestrial and freshwater situations suitable for life but are almost entirely excluded from the marine environment, perhaps because of the dominance of the crustacea. Polymorphism with radically different developmental and sexual forms is characteristic of many insect orders. Most insects are highly mobile, and functional wings are present in all groups except for a few that are very primitive or parasitic. The feeding habits of insects vary both among taxonomic groups and developmental stages of the same species. There are very few substances of animal or plant origin that some insects cannot utilize as food. Some insects such as cockroaches are highly omnivorous, while others are extremely specialized in their food choices. Bisexual reproduction is the rule among insects, but parthenogenesis and paedogenesis occur. A high degree of social organization has evolved in certain groups especially the ants, bees, and termites.

Venoms have evolved in several insect groups, both as food-getting and defensive adaptations. Those used in the former capacity are often highly specific in action and extremely potent. Venom of a small braconid wasp acts only on the voluntary muscles of lepidopteran larvae, and one milligram is enough to paralyze about 200 kg of caterpillar. Venoms used primarily for defense tend to have a wide spectrum of activity, but are lower in toxicity (Beard, 1963). Insect salivary secretions injected incident to parasitism or micropredation may be intrinsically toxic, although more frequently they are potent allergens. Besides substances injected parenterally, many insects have toxic, repugnatorial, or allergenic substances that may be deposited on the skin or mucous membranes.

Envenomation by insects is a well nigh universal human experience. Although usually only uncomfortable or annoying, it can cause serious illness and endanger life. Insect stings are the leading cause of deaths from venomous animals in the United States and probably in Europe. Insect envenomation cannot always be clearly separated from parasitism or micropredation; however, the insect groups considered here are free living and do not feed upon vertebrate blood or tissue fluids.

II. HYMENOPTERA (BEES, WASPS AND ANTS)

The social hymenoptera are by far the most important of venomous insects, for their colonies react aggressively to disturbance and can inflict hundreds of stings in a few minutes. Some of the solitary wasps are fierce stingers but do not attack in groups. The honey bee *(Apis mellifera)* because of its wide distribution and semi-domestic status is probably the leading cause of hymenopteran envenomation, but other species of bees are also of some significance. Hornets and yellowjackets *(Vespa, Vespula* and *Dolichovespula* species) and paper wasps *(Polistes)* are also abundant and widespread and are more aggressive and have a more powerful venom. Some of the ants such as the fire ants *(Solenopsis),* army ants *(Eciton),* and other tropical species attack savagely but are more easily evaded than the flying forms.

The hymenopteran venom apparatus is located in the terminal portion of the abdomen. While subject to considerable morphological variation among the various families and genera, the general features are similar. One or two tubular acid glands empty into a pyriform or fusiform venom reservoir whose duct discharges into the sting. The alkaline or Dufour's gland also empties its secretion into the base of the sting, its duct usually fusing with that of the venom reservoir. The sting is heavily chitinized in most species and retractile. In the honey bee and a few other species, the tip is barbed, so that the entire venom apparatus may

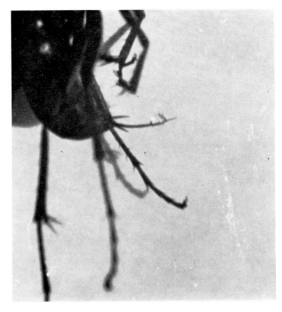

Figure 20. Posterior end of wasp's abdomen with sting extruded and drop of venom near its tip. (Photo by author.)

be torn out of the insect's body with an effective sting.

The contributions of the various secretory cells of the glands to the complex mixture that is hymenopteran venom are incompletely known. The secretory granules of the acid gland of the oriental hornet stain positively for lipids and carbohydrates; they show only traces of protein (Kanwar and Sethi, 1971). The alkaline gland may be the site of synthesis or secretion of such low molecular weight, pharmocologically active compounds as serotonin, histamine, and acetylcholine which are present in many hymenopteran venoms but absent in the venom of the mud dauber *(Sceliphron)* which lacks this gland (Rosenbrook and O'Conner, 1964a).

Until about 1960, most work on the chemistry and immunology of hymenopteran venoms was done using extracts of whole insects or venom sacs and glands. More recently, pure venom has been obtained from most of the larger forms by anaesthetizing the insect with carbon dioxide, stimulating the abdomen with electric shock or pressure, and collecting venom droplets from the sting with a capillary or in a well-slide (O'Conner et al., 1963). Benton et al., (1963) described a technique for extracting venom en masse from honey bees by electrically stimulating them to sting through a sheet of nylon fitted under the brood chamber. An entire hive can be "milked" in about five minutes and with less than 1 per cent mortality. Individual bees and wasps yield 0.01 to 0.001 ml of venom using these techniques. Volumes up to 0.1 ml can be obtained by expressing the entire content of the venom sac in the larger species. O'Conner et al. (1964a) have shown that there are significant differences in the protein composition of the venom and that of the insect's body tissues.

Honey bee venom contains approximately 88 per cent water. Between 50 and 70 per cent of the venom solids are proteins; the remaining material contains peptides, free amino acids, lipids, and carbohydrates (O'Conner et al., 1967). There is considerable variation in quantity and composition of the venom. Secretion in the worker bee begins shortly before emergence and reaches a peak between the tenth and sixteenth day after emergence. Secretion ceases after 20 days except in bees that overwinter. Protein in the form of pollen is essential for full production of venom (Beard, 1963).

Disc gel electrophoresis and immunoelectrophoresis of bee venom indicate the presence of at least eight high molecular weight fractions, five with cathode mobility and three with anode mobility. The material present in highest concentration is melittin, which may make up as much as 50 per cent of the dry weight of the venom. Although originally reported as a protein, recent work shows that it is a polypeptide composed of 12 different amino acids and having a molecular weight of 2,840. In the venom it occurs in micelles of four molecules. It has strong surface tension lowering activity, and its structure may be comparable to that of some detergents. Melittin and three synthetic cogeners have been synthesized (Lübke et al., 1971). Apamin is a second polypeptid composed of 10 different amino acids and having a molecular weight of 2,036. It is high in cystine and contains two disulfide bridges but is lacking in aromatic amino acids. It is not antigenic. The principal enzymes in bee venom are phospholipase A and hyaluronidase. Phospholipase B is also present (Habermann and Reiz, 1965; Munjal and Elliott, 1971). Pharmacologically active low molecular weight substances in the venom include histamine, dopamine, and noradrenalin (Habermann, 1972).

A pharmacologically active polypeptide with properties similar to bradykinin has been isolated from venom of the European wasp *(Vespa vulgaris)* and similar compounds from the European hornet *(V. crabro)* and paper wasps of the genus *Polistes*. These substances, known as wasp kinins, are slowly dialyzable and heat stable but inactivated by chymotrypsin. They probably occur in the venoms of other social wasps (Schachter and Thain, 1954; Bhoola et al., 1961; Pisano, 1966). Hyaluronidase activity of *Vespa vulgaris* venom is considerably higher than that of the honey bee (Jaques, 1956) and this enzyme occurs in venoms of other related wasps. Protease and phospholipase activities have been reported for wasp and hornet venoms, but the enzymes have not been adequately characterized.

Much of the toxicity of venoms of social wasps, and particularly their ability to cause pain, is associated with low molecular weight substances. Venom sacs of *Vespa crabro* have the highest content of serotonin of any natural source, and high concentrations were found in the venom sacs of other hornets, the large tropical American wasp, *Synoeca surinama*, and the paper wasps of the genus *Polistes* (Welsh and Batty, 1963). Acetylcholine is present in comparatively large amounts in hornet venoms, making up 1.8 to 5 per cent of the dry weight, and histamine is present in about half this concentration (Bhoola et al., 1961). Adrenaline, noradrenalin, and dopamine have been reported in the venom of the oriental hornet (Edery et al., 1972).

Venoms of the solitary wasps have been less thoroughly investigated than those of the social species. That of the mud dauber *(Sceliphron caementarium)* is comparatively low in protein but contains 17 low molecular weight compounds including free amino acids, lecithin, and a number of unidentified substances. It contains no kinins, serotonin, histamine, or acetylcholine. Its proteins are immunologically different from those of paper wasp, yellowjacket, and honey bee venom (Rosenbrook and O'Conner, 1964a, b). Welsh and Batty (1963) found no serotonin in the venom organs of three other species of solitary wasps.

Ant venoms and related secretions are a fascinating group of compounds. Those of the more primitive ants (Myrmecines, Ponerines, Pseudomyrmecines, and Dorylines) are similar to venoms of social wasps and bees. The red bull ant *(Myrmecia gulosa)* secretes a heat labile protein, hyaluronidase, a substance with kinin-like activity, and histamine (Cavill et al., 1964). Venoms of the army ants *(Eciton)* are evidently of this general type. Fire ant venoms are unique among the stinging arthropods in being without protein or polypeptid. The active components are alkaloids distinctive for each of the four forms in the U. S. Venom of the imported red fire ant *(Solenopsis saevissima)* was shown to contain five alkaloids, three *trans*-2 methyl-6-alkylpiperidines and two unsaturated analogues. Composition was verified by synthesis (MacConnell et al., 1971). This venom is unique in having its alkaloids in the *trans* rather than the *cis* configuration.

In the more highly-evolved Dolichoderine ants, the venom apparatus is reduced and largely replaced by anal glands which produce volatile alarm and defense secretions which are oxygenated aliphatic or terpenoid compounds of low molecular weight. In the Formicine ants, the venom glands secrete large quantities of formic acid (Cavill and Robertson, 1965).

When injected intravenously into cats and mice, whole hornet *(Vespa orientalis)* venom causes dyspnea, diarrhea, akinesia, paresis, and protracted paralysis, apparently from a combination of central and peripheral action. Pulmonary congestion and fluid in the respiratory tract are the chief findings at autopsy. The LD_{50} for mice is 2.5 mg per kilo. The venom causes contraction of the guinea pig ileum and bronchial musculature. This effect is modified but not completely abolished by histamine, serotonin, and acetylcholine antagonists; the residual smooth muscle stimulating action is presumed due to kinin. An increase in skin capillary permiability occurs. This is presumed due to histamine and serotonin since it was largely abolished by antagonists to these substances. There is degranulation of mast cells and hemolysis of blood agar. The venom has no effect on nerve conduction in muscle-nerve preparations (Edry et al., 1972). Much of the lethal effect is apparently due to nonantigenic toxins, for an antiserum produced against the four antigenic venom components had only a weak protective action (Ishay et al., 1971).

Much of the pharmacologic activity of bee venom can be ascribed to histamine. Appreciable amounts of this amine are present in a free state in the venom; in addition, it liberates histamine from mammalian tissue by three different mechanisms. Histamine from degranulation of mast cells is brought about by a polypeptide fraction designated as F II (Fredholm and Haegermark, 1967); phospholipase A releases histamine and spasmogenic lipids from guinea pig lung tissue which is unaffected by F II; melittin is also reported to be a histamine releaser (Fredholm et al., 1969). Other pharmacological activities of melittin include production of a fall in blood pressure followed by a rise, damage to motor end plates of nerves, and a lowering of the membrane potential of striated muscle. It hemolyzes washed erythrocytes and releases serotonin from platelets (Neumann and Habermann, 1956). Melittin in sublethal doses has been reported to protect mice against doses of x-irradiation fatal to 50 per cent of controls (Ginsberg et al., 1968). Melittin's surfactant properties are believed to account for much of its biological activity. Apamin is a neurotoxin with the spinal cord as its principal site of action. It produces prolonged hyperexcitability by augmenting polysynaptic reflexes (Habermann, 1972).

There is little information on the action of the venoms of solitary wasps. Against their normal prey, spiders, caterpillars, and other arthropods, the venom typically induces a prolonged paralysis of the locomotor system sometimes lasting for weeks. Vital functions such as heartbeat and digestion persist, and spiders may continue to produce silk. The metabolism is that of a starving insect. In the case of *Macrobracon hebetor* and *Philanthus triangulum,* the venom blocks neuromuscular transmission at a presynaptic site (Rathmayer, 1966; Piek and Thomas, 1969).

Pogonomyrmex barbatus, a common desert ant of the southwestern U. S., has a venom with strong cholinergic properties producing piloerection and sweating in minute doses. Fatal doses injected into mice cause dragging and twisting of the abdomen, respiratory distress, lethargy, and terminal convulsions. The LD_{50} is 24 mg per kilo (Williams and Williams, 1964) . Fire ant *(Solenopsis saevissima)* venom has potent insecticidal activity. Applied topically or as a residue, it was at least as toxic as DDT to *Drosophila* and was highly toxic to four other species of insects and two species of mites. The fire ant itself was not very susceptible. The action of the venom on arthropods is to produce rapid paralysis. The venom also has antibiotic activity against several species of bacteria and molds and causes hemolysis of rabbit erythrocytes (Blum et al., 1958; Adrouny et al., 1959) . The insecticidal activity may be related to the *trans* configuration of the molecule and the unsaturated alkenyl side chain. Venoms of two other fire ants, *S. xyloni* and *S. geminata,* with somewhat different molecular structure were less active against insects (Brand et al., 1972) .

A characteristic and clinically important property of many hymenopteran venoms is their ability to induce a high degree of allergic hypersensitivity in man. This may culminate in anaphylactic shock or other life-endangering reactions. Investigations of the antigenic composition and specificity of hymenopteran venoms indicate that several antigenic substances are present in all venoms save those of the fire ants. Some of these antigens are also present in the venom sac and body tissues of the insect but most are not. Using venom obtained by electrical stimulation, Arbesman et al., (1965) demonstrated 12 or 13 antigens in paper wasp *(Polistes)* venom, 9 or 10 in yellowjacket *(Vespula),* and 9 in honey bee. They found no antigens in common between bee and yellowjacket or between bee and paper wasp, but two antigens were shared by wasp and yellowjacket. Using similar methodology, O'Conner and Erickson (1965) found common antigens in the venoms of paper wasp, honey bee, and two species of hornets. Foubert and Stier (1958) found guinea pigs sensitized by whole-insect extracts of yellowjacket reacted with anaphylactic shock to injections of honey bee, paper wasp, black hornet and yellow hornet extracts. Animals sensitized with honey bee extracts showed little sensitivity to wasp, hornet, or yellowjacket antigens, but hornet-sensitized animals were sometimes shocked by honey bee antigen. Using passive transfer tests and *in vivo* neutralization of passive transfer in human subjects, evidence was found for a common allergen in paper wasp and yellowjacket, a different allergen common to bee and yellowjacket, and a third common to bee and paper wasp. One allergen was restricted to yellowjacket. The allergens were detected in the venom sac and sacless insect body as well as in the venom (Langlois et al., 1965).

Stinging hymenoptera occur virtually throughout the world. Some species such as the honey bee are cosmopolitan, and most of the medically important species have wide ranges and habits that bring them into close contact with man. Under natural conditions, honey bees nest in hollow trees and crevices in rocks; however, they may utilize the walls of occupied houses. Paper wasps frequently suspend their nests from window frames or under eaves or porches. The much larger paper nests of hornets may also be found in such places, although they prefer trees. Yellowjackets nest underground, often in rotted out stumps. Bumble bees also nest next to the ground frequently utilizing old mouse nests. In the southern United States, fire ant mounds are usually built in pastures of paspalum, Bermuda grass, or clover, or in corn, cotton, and sorghum fields. Ants in the tropics utilize a great variety of situations for their nests. Mud daubers frequently plaster their nests on the walls of buildings or under bridges; other solitary wasps may dig burrows.

In the western world, at least, the incidence of serious insect stings is higher in adults than in children and higher in males than in females. Multiple stings nearly always result from disturbance of an insect nest. Among honey bees, it has been observed that the act of stinging releases an alarm substance that incites aggressive behavior in other members of the colony (Benton et al., 1963). Similar substances released by ants also stir colonies to a frenzy of irritability (Cavill and Robertson, 1965). The great majority of persons are stung while engaged in outdoor work or recreation. Bee keepers are a special high risk occupational group, although many develop a high degree of immunity as a result of frequent stings. Other relatively high risk occupations are farmer, house painter, carpenter, and bulldozer operator. Wasps and bees are frequently swept into the interior of moving automobiles, an experience that frequently leaves the insect in a highly irritable state and exposes the human occupants to the risk of both a sting and a highway accident. Ripe and decaying fruit attracts yellowjackets, hornets, and paper wasps, while flowers and some perfumes attract bees.

In the temperate zones, most bee, wasp, and ant colonies either die out in the fall except for over-wintering females or are reduced in numbers. In spring, colonies build up slowly at first and reach peak numbers in the late summer and early fall. The seasonal incidence of stings in man reflects this cycle.

Aside from the deliberate introduction of the honey bee, there have been at least two important introductions of stinging insects into the United States. The best known of these is that of the tropical American fire ant, which was accidentally imported into the area of Mobile, Alabama about 1920. It now occurs in at least 10 southern states and has largely displaced the two native species of fire ants in this region. The European hornet *(Vespa crabro)* was introduced into the eastern states and is now widely distributed in the eastern half of the country. It is a larger insect than the native hornet *(Dolichovespula maculata),* forms larger colonies, and shows a greater tendency to nest close to human habitation.

Single wasp, bee, or ant stings in unsensitized individuals cause instant pain accompanied by a wheal and flare reaction and a variable amount of edema. With the more venomous species, the pain and swelling may extend well beyond the site of the sting and persist for as long as four days. Multiple stings, particularly by such species as the giant bee of India *(Apis dorsata),* the Guinea wasp *(Synoeca surinama)* and the larger hornets may be followed by vomiting, diarrhea, generalized edema, shock, and renal damage. Widespread necrosis of skeletal muscle, hyperkalemia, and tubular necrosis of the kidney have been reported following envenomation by the large hornet, *Vespa affinis.* Adults have died after about twenty stings (Scragg and Szent-Ivany, 1965; Shilkin

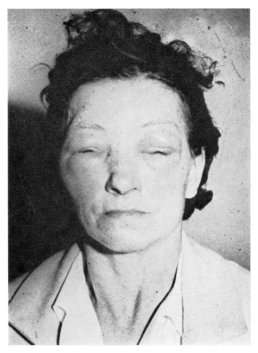

Figure 21. Patient with facial edema following multiple bee stings. Photographed 6 hours after injury. (Photo courtesy Z. Maretic.)

et al., 1972). I have known renal shut-down to follow multiple stings by the white-faced hornet *(Dolichovespula maculata)*. A fatality preceded by convulsions and complete hemiplegia occurred 30 hours after a patient suffered about 60 yellowjacket stings. Autopsy showed massive cerebral infarction and marked pulmonary congestion (Day, 1962).

In the United States and Europe, death from anaphylactic reactions to stings is much more frequent than death from toxemia. Of 229 insect sting fatalities between 1950 and 1959, only eight were ascribed to envenomation (Parrish, 1963). Serious but nonfatal allergic reactions are comparatively common. Although the insects involved are frequently unidentified or inadequately identified, there is reason to think the yellowjacket figures in an unusually large number of cases, with the honey bee and paper wasp next in order. Anaphylactoid reactions including at least one fatality have followed fire ant stings (Caro et al., 1957; Brown, 1972). Since the toxin of these ants is not a complete antigen, it either acts as a haptene in inducing hypersensitivity or it triggers the release of autopharmacological substances whose effects simulate the picture of anaphylaxis. Stings by other unidentified ants have been followed by anaphylactoid reactions (Morehouse, 1949). Venoms whose intrinsic toxic effect on man is mild such as that of the mud dauber may induce hypersensitivity and trigger fatal shock (O'Conner et al., 1964b).

Insect sting anaphylaxis does not differ in its clinical manifestations from that triggered by other substances. It may follow a single sting, and nearly always manifests itself within 40 minutes. Milder reactions of hypersensitivity such as hives, edema, nausea, and wheezing are sometimes seen and are warning of a dangerously allergic state. In fatal cases, the autopsy findings include cyanosis, pulmonary and laryngeal edema, mucus in the air passages, visceral congestion, and acute cardiac dilitation. For case histories and additional clinical and pathological details see Perlman (1955), Mann and Bates (1960), Jensen (1962), and Barnard (1967). Barnard (1966) has reported cases of delayed fatalities occurring from 8 to 109 days after single or multiple hymenopteran stings. There was no common theme in either the clinical picture or autopsy findings. Multiple internal hemorrhages, central nervous system necrosis, laryngeal edema and emphysema, and cardiac infarction were among the pathological manifestations. The pathogenesis remains obscure.

The identification of the individual with potentially dangerous allergy to insects is not always possible. Mueller (1959) reported that in 63 of 84 patients with severe reactions to stings, there was a personal or family history of allergy. On the other hand, in an analysis of 99 fatal stings O'Conner et al. (1964b) obtained a history of allergy in only three. However, in 17 of 36 cases where information was available, there was a history of previous severe or atypical reaction to insect sting. Skin tests, passive cutaneous anaphylaxis, and other immunological techniques sometimes give equivocal or contradictory results (Shulman,

1968). McCormick (1963) demonstrated precipitins to wasp and bee venom in the serum of a man who died less than 15 minutes after being stung by a paper wasp. This technique is of potential medicolegal importance in cases of unexplained sudden death.

Local reactions to most hymenopteran stings can be controlled by rest, elevation of the extremity, application of ice packs, and administration of one of the rapidly acting antihistaminics (tripelenamine or diphenhydramine hydrochloride). As might be expected, antihistaminics are of little help in fire ant stings. Severe systemic reactions, whether anaphylactic or toxic, should be treated with epinephrine and parenteral corticosteroids. Shock and renal failure should be treated by standard methods. An open airway must be maintained and oxygen administered. Milder reactions usually respond to isoproterenol and antihistaminics (Perlman, 1962; Bladek, 1968).

Persons known to be sensitive to insect stings should carry a kit with emergency drugs and equipment for their administration. Desensitization is advisable but should be carried out under supervision of a specialist, for it is sometimes impossible to design a safe regimen for the patient hypersensitive to insect venom. Some clinicians prefer to use whole-body extracts of insects believing that antigens other than those contained in the venom may play a part in hypersensitivity (Perlman, 1962; Frazier, 1964). Combination of venom with adjuvant and use of periodic natural stings to sustain immunity has been reported to maintain hyposensitization over a prolonged period (Loveless, 1962; 1968).

As a group the hymenoptera are highly susceptible to most insecticide sprays, and their control around dwellings and other inhabited buildings is rarely difficult. Because many species are economically valuable as pollinators of plants or predators on other insects, their control on a wider scale is undesirable.

III. LEPIDOPTERA

Another common but less serious type of insect envenomation is that caused by lepidoptera. It most commonly results from contact with caterpillars and less frequently with the cocoon or adult stage. Many species have been incriminated. Some of the most severe lepidopteran envenomations have been attributed to larvae of the genus *Megalopyge* which occur from the southern United States through much of tropical America. The saddleback caterpillar *(Sibine stimulea)*, tree asp *(Euclea delphinii)*, and the large

Figure 22. Pine Caterpillars *(Thaumatopoea pyrocampa)* in typical processionary arrangement. This is a common species in Europe and the Mediterranean region and a frequent cause of dermatitis. (Photo by Z. Maretic.)

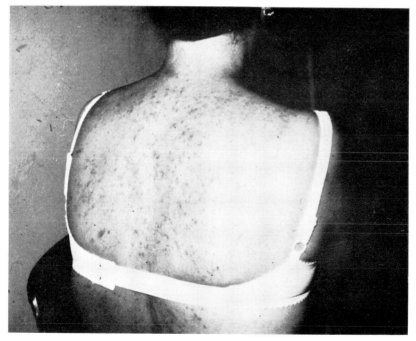

Figure 23. Maculopapular skin lesions caused by contact with detached spines of Pine Caterpillars. (Photo courtesy Z. Maretic.)

and showy caterpillar of the Io moth *(Automeris io)* are well known species in the eastern United States. Pine caterpillars *(Thaumatopoea)* and processionary caterpillars *(Ctenocampa)* are important in Europe and the Mediterranean region, while the browntail moth caterpillar *(Nygmia phaeorrhoea)* is troublesome both in Europe and in the U. S. where it has been introduced. Female moths of the genus *Hylesia* have irritant spines. They are native to South America and swarm in enormous numbers during the rainy season at which time they are attracted to lights. Moths and caterpillars of the genus *Euproctus* are the most frequently involved species in southeast Asia and the Far East.

Lepidopteran venoms are entirely defensive in function, and the venom apparatus is simple in the few species that have been well investigated. Moths of several genera, such as *Anaphe* and *Hylesia,* have hollow, barbed spines around the tip of the abdomen. These fit loosely into cups containing glands that secrete the venom which is drawn into the lumen of the spine by capillary action. One of the hawk moths *(Acanthosphinx guessfeldtii)* has similar spines on its legs (Rothschild et al., 1970). In many caterpillars the spines are branched and occur in groups and tufts on the dorsum, sometimes with longer, nonvenomous spines and hairs. In other species, the venomous spines are diffusely distributed over the dorsal surface. In *Megalopyge* the base of each spine contains a bulb continuous with the lumen of the spine and presumably the site of venom secretion; in *Automeris* the venom glands are in the papillae which bear the tufts of spines. Venom fills the lumen of the spines; in some cases it may be on the surface as well. No muscles or other special structures for expelling the venom have been described.

It is difficult to obtain caterpillar venom in pure form. This has been a drawback in studying its chemistry and biological activity. Early work with *Megalopyge opercularis* demonstrated that the ability of the spines to cause epidermal necrosis was abolished by 10 minutes boiling or heating overnight at $55°$ (Foot, 1922). Investigation of venoms of several North American caterpillars failed to reveal the presence of either histamine or serotonin. The heat labile nature of the toxins suggested that they were proteins or polypeptides (Goldman et al., 1960). Spines of the South American tatorana or fire caterpillar (*Dirphia* sp.) were found to contain 0.02 to 0.04 per cent histamine, but only traces were found in neotropical species of *Megalopyge*. Both species were negative for acetylcholine (Valle et al., 1954). A large amount of serotonin was found in spines of a Brazilian species of *Automeris* (Welsh and Batty, 1963).

Venom of *Megalopyge urens,* a South American caterpillar known to inflict severe stings, contained proteins with many of the properties of globulins. The venom had proteolytic and hyaluronidase activity and a hemolysin for dog and human erythrocytes. Determinations for several other enzymes, histamine, acetylcholine, and factors affecting the clotting mechanism were negative. The venom was antigenic and stable for one to two weeks in the refrigerator but was inactivated by drying and lyophilization (Ardao et al., 1966).

Setae of *Automeris io* applied to the skin of animals caused rapid development of an erythematous wheal with edema similar to the action of histamine. Extracts of setae of *Megalopyge* and *Dirphia* caused a marked fall in blood pressure in dogs leading to fatal shock. This was believed greater than could be accounted for by the histamine content of the material (Jones and Miller, 1959; Valle et al., 1954). Secretion from spines of an unnamed South American saturnid caterpillar was found to have fibrinolytic activity and to activate human but not bovine plasminogen. These activities were abolished by heating to 70° and by soybean trypsin inhibitor, E-aminocaproic acid, and aprotinin (Arocha-Pinango and Layrisse, 1969).

Cases of caterpillar dermatitis have been reported from most parts of the world. Severe outbreaks due to *Megalopyge opercularis* involving more than two thousand individuals have been reported from southeastern Texas and smaller outbreaks from other southern states. These occur irregularly apparently when local ecological conditions favor increase of the caterpillars and inhibition of the parasitic wasps and diptera that are their chief natural control. Incidence of stings is highest from late August through early November (Micks, 1952; McGovern et al., 1961). In tropical

South America, dermatitis caused by *Hylesia* moths reaches a peak during the principal rainy season (April through June) with secondary peaks during December and January and occasionally in August and September. Caterpillar envenomation chiefly caused by three species of *Megalopyge* and one of *Podalia* is more generally distributed throughout the year, but again there tend to be peaks coinciding with the rainy season. About 77 per cent of accidents involve agricultural workers (Pesce and Delgado, 1966).

Contact with spines of *Megalopyge, Dirphia, Automeris* and related species causes violent pain that tends to radiate centrally. Headache, nausea, vomiting, and lymphadenitis are not uncommon. Shock and convulsions have been reported (Goldman et al., 1960; McGovern et al., 1961; Pesce and Delgado, 1966). Most of these effects disappear within 24 hours. Five patients stung by an unidentified saturnid caterpillar in Venezuela developed a hemorrhagic syndrome 8 to 72 hours after injury. This was manifest by petechae, melena, and bleeding from the nose, ears, and vagina. Fibrinogen levels were decreased but platelets were normal. Up to five weeks were required for recovery, but this was shortened by administration of fibrinogen (Arocha-Pinango and Layrisse, 1969). With some other species the effects are more delayed and take the form of an itching papular dermatitis or a generalized urticarial eruption that may persist several days. The spines of some species of caterpillars and moths are readily detached and retain their irritative properties. In Israel some 600 cases of dermatitis occurred among soldiers camped in a pine grove. No live caterpillars were found, but dead caterpillars, cocoons, and debris from under the trees produced dermatitis in volunteers (Ziprkowski et al., 1959). *Hylesia* dermatitis is usually acquired by contact with detached spines that become widely distributed during periods when the moths are plentiful. Thirty-one of 45 members of a ship's crew exposed in a Venezuelan port developed dermatitis, most of them from sleeping on bed sheets on which the moths had alighted (Hill et al., 1948).

Detached spines lodging in the eyes can cause painful ophthalmia. In a few cases where large numbers of spines have become embedded in the eye, intense and intractable endophthalmitis has resulted and necessitated removal of the affected eye. Spines have been found in ocular tissues as much as five months after the original injury (Gundersen et al., 1950; Dreyer, 1953; Corkey, 1955).

Local treatment has little effect on stings by *Megalopyge* and other more toxic caterpillars, although prompt application of adhesive or scotch tape will sometimes remove many spines. Intravenous calcium gluconate has been reported to be quite effective in severe envenomation, and analgesics such as meperidine will relieve the pain. Some cases of cutaneous envenomation have responded well to lotions and creams containing steroids. Antihistaminics are of little value. Wide individual variation in response to lepidopteran stings occurs suggesting that hypersensitivity plays a role. Reactions of the anaphylactic type have not been reported, however.

Insecticidal spraying of host plants is usually adequate to control dermatitis-producing caterpillars in thickly populated areas. Screening protects against moth invasions.

IV. HEMIPTERA

A number of the predacious hemiptera are capable of inflicting venomous bites. Most of these insects use their venom to subdue the small creatures on which they feed and they bite man only as a defensive response. A few hemiptera such as the triatomids and cimicids are micropredators on vertebrates including man, and severe reactions of an allergic nature have been reported to their bites.

The hemipteran venom apparatus consists of two pairs of salivary glands, the larger lying in the mesothorax next to the proventriculus and the smaller accessory glands posterior to them. Their mixed secretions are ejected through half of a double tube formed by the interlocking of the very elongate and highly modified maxillae and mandibles whose distal tips are modified as piercing lancets. These in turn are enclosed in a segmented sheath, the whole forming a beak which is prominent in many species. In feeding, the insect inserts its interlocked mouthparts, injects a powerfully lytic saliva, and simultaneously sucks the body fluids and liquified tissues of its prey.

The only hemipteran venom to receive much study is that of *Platymeris rhadamanthus,* a large African assassin or reduviid bug. This material is readily obtained because of the bug's spitting or squirting out its venom as a defensive response. The dried venom so obtained is stable for at least three years. It has at least six protein fractions and contains endopeptidase, hyaluronidase, and a weak phospholipase. Lipase, esterase, ATPase, and serotonin were not detected. The venom of another large reduviid, *Holotrichius innesi* of the Middle East, contains 13 protein fractions and has gelatinase, esterase, and hyaluronidase. It has no phospholipase but does have a direct hemolysin for mammalian erythrocytes.

In insects, *Platymeris* venom abolished cardiac contractions and caused intense neuroelectrical activity followed by cessation of neural conduction. Insects were paralyzed in 3 to 5 seconds. A general lysis of all but cuticular and collagenous structures followed. *Holotrichius* venom had similar effects. Natural bites by this insect killed mice in 15-30 seconds, and the intravenous LD_{50} was established at 1 mg/kg. Convulsions, respiratory paralysis, and internal hemorrhages were seen. Venom aerosols were fatal to guinea pigs (Edwards, 1961; Zerachia et al., 1972) .

Salivary gland homogenates from other assassin bugs *(Rhinocoris carmelita* and *Reduvius personatus)* and a water bug *(Naucoris cimicoides)* were about as toxic as that of *Platymeris;* however, homogenates from *Rhodinus* and *Triatoma* that feed on mammalian blood were nontoxic for insects. *Platymeris* hemolymph was not toxic (Edwards, 1961) .

Venom of the giant tropical American waterbug, *Lethocerus del-ponti,* was toxic for guinea pigs, frogs, and small fish producing hemorrhages and paralysis. It liquified gelatin and hastened the coagulation of rabbit and human blood. It was strongly hemolytic for rabbit erythrocytes but not those of man (Picado, 1936) .

There are few accounts of human injuries by nonbloodsucking hemiptera, although entomologists who collect assassin bugs soon learn that a certain degree of caution is advisable. I vividly recall the painful stab that the black corsair *(Melanolestes picipes)* can inflict when picked up. *Platymeris* bites can cause local pain,

Figure 24. Wheel-bug *(Arilus cristatus)* This large predatory insect is common in the eastern United States during late summer. It can inflict a painful and mildly venomous stab wound if molested. (Photo by author.)

swelling, and necrosis, and the dried venom is irritant to the eyes and nasal mucous membranes (Edwards, 1961). The wheel-bug *(Arilus cristatus)* which is the largest reduviid in the U. S., has been reported to have caused cases of envenomation in Maryland and Nebraska. A child of 10 months was bitten on the scalp while in his home. Pain and local swelling lasting less than 12 hours were the only manifestations (Smith et al., 1958). An older child bitten on the finger developed two papillomatous lesions near the bite; these persisted six to nine months. Entomological literature from the early part of the century contains reports of cases with systemic manifestations such as fever, hives, and nausea attributed to *Reduvius personatus* (Hall, 1924).

Several types of aquatic hemiptera such as the back-swimmers (Notonectide), water scorpions (Nepidae), and giant waterbugs (Belostomatidae) can inflict venomous bites. These insects prefer quiet, warm, shallow water but have been found in many aquatic situations including swimming pools. The giant waterbugs are readily attracted to lights. All these insects may bite when handled or if trapped under clothing. Bites inflicted by *Benacus griseus,* a common species of the eastern U. S., on volunteers were described as "rather painful and more than transitory" producing redness and swelling lasting about four hours (Ewing, 1928). The larger tropical American *Lethocerus del-ponti* can give a bite that has been likened to bites by the smaller pit vipers of that region (Picado, 1936).

V. COLEOPTERA (BEETLES)

In numbers of species, beetles are the largest group of insects, but their venomous properties as far as man is concerned are virtually confined to the elaboration of irritant secretions. A few species such as the stag beetles can inflict a painful pinch with their enlarged mandibles, but there is no associated venom. The predacious aquatic larvae of the large diving beetles have powerful, sickle-shaped canulated mandibles. Although venom has not been demonstrated, the speed with which they kill small fish and tadpoles suggests the presence of some toxic substance. No proved instance of injury to man has come to my attention; however, I consider it entirely possible.

The blister beetles of the family Meloidae are a widely distributed group most characteristic of semiarid and desert regions. A species may suddenly appear by the thousands, especially after rains, and then disappear in a few days only to be replaced by another. The beetles are plentiful on vegetation and some species are attracted to lights. Some are brilliantly colored and attractive to children. The toxin of all blister beetles is cantharidin ($C_{10}H_{12}O_4$) which is present in the hemolymph and most of the insects' tissues, sometimes making up as much as 2.5 per cent of the dry weight. It exudes from multiple sites if the beetle is gently pressed or otherwise injured. Cantharidin has a strong vesicant effect, although its activity is confined to the outermost layers of the skin (Lehmann et al., 1955). It is also a powerful irritant to the urinary and gastrointestinal tracts.

Blister beetle dermatitis is not an uncommon condition. In the eastern United States it is usually caused by *Epicauta cinerea* or related species. Contact with the beetle is painless and seldom remembered by the patient. The blisters usually develop two to five hours after contact and may be single or multiple. They are ordinarily 5 to 50 mm in diameter, thin walled, and often linear in arrangement. Unless broken and rubbed, they are not painful. Cantharidin nephritis has been reported following unusually severe vesication; however, it is more frequently the result of using cantharidin preparations as an aphrodisiac (Swarts and Wanamaker, 1946; Theodorides, 1954; Browne, 1960).

Another type of beetle vesication is caused by small rove beetles of the genus *Paederus* which occur in many temperate and tropical regions but not in the United States. They live in damp, forested habitats and fly during the evening. Their activity is usually confined to hot, humid weather. The vesicant substance produced by these beetles is chemically distinct from cantharidin. It is present in greatest concentration in the haemolymph but does not exude spontaneously as in the blister beetles. Vesication occurs if the beetles are crushed or rubbed on the skin. The first manifestation is erythema which begins about ten hours after contact. Forty-eight hours later a crop of small blisters appear and persist for about two days. This is followed by desquamation and healing. There is some geographic variation in the severity of symptoms. An East African species is responsible for an acute conjunctivitis known as "Nairobi eye" (Theodorides, 1950; Fain, 1966).

There is no specific treatment for beetle vesication. Large blisters should be drained and the site covered with a protective dressing. Local applications of corticosteroids or antihistaminics are not particularly effective (Lehmann et al., 1955; Fain, 1966). In cases of systemic cantharidin poisoning from ingestion, oils and emetics other than apomorphine are contraindicated. The acute nephritis usually responds to conventional, conservative management.

The darkling ground beetles (Tenebrionidae) are characteristic of arid regions where they live under stones and other cover and crawl about at night. They have irritant secretions, mostly benzoquinones, that may be sprayed 30-40 cm from the tip of the abdomen. Large species such as *Blaps nitens* and *B. judaeorum* of the Middle East can cause blistering of the skin.

There are unquestionably other types of rare and sporadic insect envenomation not adequately described in medical literature. Years ago in Florida I was told that the stick insect or "seven-day rider" *(Anisomorpha buprestoides)* could blind a man with its poisonous spray. This insect does eject from its thoracic region a noxious fluid that deters birds and other predators (Eisner and Meinwald, 1966) and might well do damage to human eyes.

Chapter 4

VENOMOUS FISHES

Fishes are the most numerous of vertebrates with some 40 thousand species inhabiting nearly all aquatic environments. The smallest species are not much more than 10 mm long when adult; the largest reach lengths of at least 15 meters. The vast majority of living species belong to the Osteichthyes or bony fishes; however, the Chondrichthyes with a largely cartilaginous skeleton include some very large and medically important species among the sharks and rays.

Considering the variety of fishes and the numerous ecological niches they occupy, it is somewhat surprising that venom has been identified in only some 200 species. The venomous species belong to nine families of cartilaginous fishes and at least 17 families of bony fishes. In nearly all species the venom apparatus functions only as a defensive adaptation and is associated with spines on various parts of the body, usually the dorsal fin, tail, operculum, or pectoral region. Recently it has been shown that the small blennies of the genus *Meiacanthus* have a pair of deeply grooved teeth in the lower jaw associated with what appear to be venom glands. Bites of *M. nigrolineatus* on a human volunteer produced a wheal and flare type response followed by induration lasting about 24 hours. No such response followed bites by the very similar *Plagiotremus townsendi*. Experiments with potential predators indicated *M. nigrolineatus* was a generally unacceptable prey species because of its bite (Springer and Smith-Vaniz, 1972). Although of no known medical significance, the discovery in fish of this type venom apparatus, hitherto associated with lizards and insectivores, is of considerable biological interest.

I. STINGRAYS, SHARKS, AND CHIMERAS

The stingrays are a major group of venomous fishes. They are markedly flattened and circular to rhomboidal in shape. The tail may be extremely long and whip-like or very short; the venomous spine may be located near the middle of the tail or at about the juncture of the proximal third of the tail with the distal two-thirds. The venom apparatus is best developed in the dasyatid, urolophid, and potamotrygonid rays. It is less well developed in the myliobatids and rhinopterids and poorly developed in the gymnurids. Most of the mobulids including the giant manta which may reach a width of more than 5 m are completely without a sting.

Stingrays are found throughout the world in tropical and warm temperate coastal waters. A few are pelagic and may be found in deep water, but the great majority inhabit shallow water with a mud or sand bottom. They characteristically bury themselves leaving only the eyes, spiracles, and part of the tail exposed. River mouths are a favorite habitat for some species. Rays of the family Potamotrygonidae are entirely freshwater and inhabit several of the major river systems of South America, the Banoue River in central Africa, and a section of the Mekong River in Laos (Castex, 1967). Most stingrays feed on mollusks, worms, and other bottom-dwelling invertebrates. The female ray retains her eggs throughout the developmental period; the young are born fully formed and have been reported as trying to sting while still in the embryonic membranes (Gudger, 1943).

The sting of the ray consists of a dentinal spine with serrate edges and two ventrolateral longitudinal grooves. In the small California round stingray *(Urolophus halleri)* the sting is about 4 cm long; in the giant stingray *(Dasyatis brevicaudata)* it may measure 37 cm. In some species it is not unusual for one animal to have two to four spines. Most of the venom secreting tissue lies in the ventrolateral grooves which are filled with glandular epithelial cells. The entire sting is encased in an integumentary sheath which also contains glandular cells. Secretion is of the holocrine type, and there is no duct system (Halstead et al., 1955; Halstead, 1970). The stinging response is elicited when the ray is pinned against the bottom or some other solid object; there is no reliable report of a ray stinging while swimming. The tail is usually whipped forward and downward, but a lateral blow can also be delivered by some species. This action drives the spine into the tissues. As it is withdrawn, the serrations produce a lacerated wound within which some of the venom-laden tissue from the sting is deposited. Those who have collected large numbers of stingrays have observed that the integumentary sheath is partly or wholly torn away in as many as 30 percent of specimens (Russell, 1965). This obviously influences the severity of envenomation. The spine may be torn free from the tail and left in the wound; however, this is the exception in human injuries. Stingray spines are sometimes found in the tissues of sharks and other predatory fish.

Pure preparations of stingray venom are difficult to obtain without contamination with other substances present in the integumentary and venom-producing tissue. Russell and coworkers (1958b) found aqueous extracts from fresh or refrigerated stings had a pH of 6.76. Toxicity of the extract was lost in 1 minute at 80° and in 18 hours at 26 degrees. In glycerol at −20°, toxicity was retained for 11 months. Indirect evidence suggested the toxins were proteins. Disc electrophoresis showed the presence of 15 fractions in tissue extracts from the stings of *Urolophus halleri* and ten fractions in extracts of sponges stabbed with fresh stings. The crude extracts contained serotonin, 5-nucleotidase, and phosphodiesterase but not phospholipase or protease. Sephadex gel filtration permitted separation of a lethal fraction with an estimated molecular weight in excess of 100,000; it showed two or three bands on disc electrophoresis (Russell, 1965).

A water-insoluble, thermostable, nonlipidic substance isolated from dried and acetone-preserved stings of *Potamotrygon motoro* of Brazilian rivers showed cholinergic activity on isolated guinea pig ileum and produced hypotension in the rat (Rodrigues, 1972). No other biologically-active material was detected, however the manner of collection and preservation would probably have denatured toxic proteins.

Mammals injected with lethal doses of stingray venom show

hyperkinesis followed by prostration and atonia with hypoactive or absent reflexes. Convulsions are occasionally seen. Peripheral vasoconstriction manifests itself in blanching of the ears and retina. Salivation, vomiting, urination, and defecation occur frequently. Respirations are slow and labored (Russell et al., 1958b). The heart is severely affected. Small doses of venom cause bradycardia with an increase of the PR interval and various degrees of atrioventricular block. The blood pressure is not markedly changed. Larger doses produce second or third degree atrioventricular block usually followed by sinus arrest. Changes in the electrocardiogram indicative of ischemia are seen. The blood pressure decreases markedly and cardiac arrhythmias terminating in irreversible standstill may occur. If the dose is sublethal, the cardiovascular changes soon return to normal (Russell and van Harreveld, 1954; Russell et al., 1957). The venom has no effect on neuromuscular transmission and little effect on the central nervous system, although the respiratory depression that is seen is believed largely due to the action of the venom on the respiratory center. The intravenous lethal dose of lyophilized crude venom for mice is estimated at 28 mg per kilo (Russell, et al., 1958a).

Stingray injuries are frequent in coastal waters throughout the tropical and warm temperate parts of the world including Australia, the Mediterranean region, and Oceania. Along coasts of the United States during a five-year period there were 1097 stingray injuries reported, of which 62 were hospitalized and two died (Russell, 1959). Stingray injuries are also seen in considerable numbers along the Parana river system in South America. The annual incidence near Asuncion, Paraguay is about two hundred cases, and considerable numbers are also reported from other localities in Paraguay, Uruguay, and Argentina (Castex, 1965). Stings are most often inflicted on persons wading in the water. In the United States, Europe, and Australia, most of those injured are bathers and swimmers. In South America and Oceania, fishermen and persons gathering other types of food in shallow water are the main group at risk. The growth of longline fishing operations on the high seas has led to exposure of fishermen to the pelagic stingrays such as *Dasyatis violacea* and *Himantura schmardae*. Two fatal injuries by these rays have been reported (Rathjen and Halstead, 1969).

The great majority of stingray injuries are on the lower extremity below the knee and result from stepping on the fish. Wounds on other parts of the body may be inflicted by rays caught in nets or on hooks or when a swimmer or diver grazes a ray buried on the bottom. The wound may be a puncture or a ragged laceration up to 18 cm long. It usually bleeds freely at first but characteristically becomes surrounded by a white or bluish white ischemic zone after a few minutes. Pain at the moment of injury is severe and tends to increase during the next 90 minutes. Weakness, fainting, and nausea probably incident to the severe pain occur frequently. Moderate edema of the injured limb is commonly observed; occasionally there are muscle contractures. Symptoms probably resulting from systemic absorption of venom include salivation, vomiting and diarrhea, muscle cramps and fasiculations, dyspnea, cardiac arrythmias, and convulsions. These are infrequent and indicative of severe envenomation. Necrosis, ulceration, and dermatitis around the site of the sting are not uncommon particularly in those stings caused by the freshwater rays, *Potamotrygon*. This evidently results from local vasoconstriction and ischemia often complicated by bacterial infection. Illustrative case histories have been reported by Halstead and Bunker (1953), Russell (1953), and Castex (1965). Most fatal cases have been associated with penetration of the thoracic or abdominal cavities or were caused by tetanus and other bacterial infections. In some cases such as the one reported by Wright-Smith (1945), death resulted from perforation of the heart; however, the two fatal cases reported by Russell et al., (1958a) were clearly due to toxemia incident to penetration of the abdominal cavity by a sting. Of considerable interest are reports of a cardiac wound and a puncture wound of the liver that were not accompanied by toxemia, and the patients made rapid recoveries after surgical repair (Ronka and Roe, 1945; Cadzow, 1960). Apparently the spines of the rays causing these injuries had lost all or most of their venom secreting tissue.

First aid for sting ray injuries consists of prompt washing of the wound with seawater or any other water available and removal of any fragments of the grayish or pinkish venom-containing tissue that can be seen in the wound. This is followed by soaking the injured area in water as hot as can be tolerated for 30 to 90 minutes. Following this, a more complete debridement can be carried out and the wound sutured. Pain can generally be relieved by meperidine or similar analgesics. Elevation of the injured part is advisable. Tetanus prophylaxis should be given, and prophylactic administration of a broad spectrum antibiotic for 48 to 72 hours may be advisable in some cases. With this management, the wounds heal quickly, and necrosis and infection are rare (Bitseff et al., 1970).

Wounds penetrating the thoracic or abdominal cavity should be surgically explored paying particular attention to removal of the fragments of the integumentary sheath and other venom-containing tissue carried in by the sting. Oxygen should be given, and shock and convulsions controlled by conventional management. There is no specific antidote for stingray venom.

Diagnosis of stingray envenomation rarely is difficult. The ragged, freely-bleeding wound is larger than that caused by most other venomous marine creatures, and the ray is often seen by the victim. Lack of progressively increasing pain differentiates injuries caused by broken glass or sharp metal objects.

A few species of sharks including the familiar spiny dogfish *(Squalus acanthias)* and at least three species of chimaeras or rat-fish have been reported to inflict venomous wounds with their dorsal spines. The venom apparatus is much like that of the stingrays; however, the spines are not markedly serrate or otherwise modified, and the venom producing tissue is less in amount. This tissue is toxic for mice (Halstead and Bunker, 1952) but the chemistry and pharmacology of the venom is unknown. Instances of human envenomation are known and summarized by Halstead (1970) and Southcott (1970). The stings are painful but less so than those of stingrays, and there are no well documented reports of serious envenomation.

II. BONY FISHES

In nearly all the teleost or bony fishes the venom apparatus is associated with spines, usually those of the fins. The venom is produced in aggregations of unicellular glands that must be ruptured in order for their contents to be discharged. The glandular cells may be partly or completely enclosed in a connective tissue capsul, and there are supporting cells; however, there are no acini or ducts. The grooves of the spines facilitate entry of the venom into the wounds they make. The spine and its associated glandular tissue are enclosed in an integumentary sheath. Penetration of the spine into the tissues of an adversary forces venom out of the glandular tissue, and it flows along the channels of the spine into the wound.

A. Scorpionfishes

The most dangerous of the venomous bony fishes belong to the family Scorpaenidae which is almost cosmopolitan in distribution although better represented in temperate and subtropical seas. A few species such as the bullrout *(Notesthes robusta)* of Australia may enter fresh water. There are several hundred species of which 57 are known to be venomous. Most of the dangerous species occur around coral reefs or kelp beds. They are sedentary, relying on their protective coloring to conceal them. They may bury themselves in sand. They feed on other fishes and invertebrates. Most species bear living young.

Venom producing tissue is present on the dorsal, anal, and pelvic spines. Halstead (1970) recognizes three types of venom apparatus in this group. In the first typified by the lionfish *(Pterois)* the spines are long and slender, the integumentary sheath thin, and the venom glands small but well developed. In the second typified by *Scorpaena,* the scorpionfish, the spines are moderately long and heavy, the integumentary sheath thick, and the venom glands of moderate size. The most dangerous species typified by the stonefish *(Synanceja),* have short and stout spines, the integumentary sheath is very thick, and the venom glands very large and well developed. A venom duct is said to be present in

Figure 25. A lionfish *(Pterois antennata)* photographed in shallow water near a coral reef. These showy fishes have highly venomous dorsal spines. They are popular aquarium species, and several recent reports of injuries have involved specimens in captivity. (Photo by author.)

Figure 26. A Scorpionfish *(Scorpaenopsis* sp.) collected at Ashmore Reefs, Timor Sea. These fish often spread their pectoral fins in a threat display when alarmed. The dorsal spines can inflict a painful injury sometimes accompanied by symptoms of systemic poisoning. (Photo by author.)

this group; however, this appears to be only the attenuated distal end of the venom gland (Cameron and Endean, 1972).

Most species sting only when touched, stepped upon, or otherwise restrained, but some adopt a more active defense. In midwater, the lionfish assumes a characteristic defensive stance that facilitates the use of its dorsal spines (Steinitz, 1959). Some of the scorpionfishes such as *Scorpaena* and *Dactylopterus* spread their brightly marked pectoral fins in a warning gesture when approached. They may follow this with lunging and butting (Breder, 1963).

Available evidence indicates a strong similarity among venoms of the Scorpaenid fishes, although only a few species have been investigated. Venom is usually obtained by extracting the macerated glandular tissue but in the stonefishes, *Synanceja,* it can be obtained by puncturing the relatively large venom glands. The yield from S. *horrida* is about 0.015 ml of fluid venom per spine or about 0.2 ml for a fish of average size. The dry weight may be as much as 130 mg. Venon may also be obtained by aspiration from the spine grooves with a micropipette. Scorpaenid venoms are clear to opalescent liquids of approximately neutral pH. Toxicity is lost in less than 30 minutes at $50°$ and in 14 to 21 days under ordinary refrigeration. Addition of Cleland's reagent tends to increase the stability of scorpionfish *(Scorpaena guttata)* venom when refrigerated (Schaeffer et al., 1971). Stonefish *(Synanceja)* venom is stable for long periods if lyophilized or stored in glycerine-saline at $-20°$ (Saunders, 1960).

The lethal property of *Scorpaena guttata* venom is associated with one or more proteins with molecular weights of 50,000 to 800,000. Rapid inactivation by parachloromercuribenzoate suggests that sulfhydryl groups are essential for toxicity. The intravenous mouse LD_{50} of the lethal fraction is 0.9 mg protein per kg; that of the crude venom 2.6 mg per kg. The crude venom contains five or six additional protein fractions that show no toxicity for mice (Schaeffer et al., 1971). Stonefish venoms contain seven or eight protein components, only one of which is toxic for mice. The intravenous mouse LD_{50} of the crude venom is 0.2 mg protein per kg; the lethal fraction is about twice as toxic (Saunders and Tokes, 1961). The toxic protein has been purified approximately ten times by starch gel electrophoresis but is quite unstable and has not been characterized (Deakins and Saunders, 1967).

In the intact animal, lethal doses of stonefish venom produce a fall in blood pressure and an increase in respiration followed by respiratory and cardiac arrest. Electrocardiograms show conduction defects and evidence of myocardial ischemia. There are tremors, muscular weakness, and convulsions. Venoms of *Pterois* and *Scorpaena* are similar in effects although of lower toxicity (Russell, 1967). Austin and colleagues (1961) have suggested that the principal pharmacological effect of *Synanceja* venom is to produce a conduction block due to slow depolarization of muscle. The paralysis affects smooth, striated, and cardiac muscle alike. Death is due to direct action on the heart and diaphragm plus loss of muscle tone in the peripheral vasculature. The pain-producing component of the venom is unknown but is evidently a protein.

Over a wide area in the Indian and south Pacific Oceans, stonefish represent a real hazard to those engaged in work or recreation that involves wading in waters around coral reefs. The spines of these fish are stout enough to pierce many kinds of footwear. No figures on the incidence of stings are available; however, Smith (1957) implies that they are fairly common in the western Indian Ocean, and an appreciable number are fatal. The colorful and distinctive lionfish and zebrafish *(Pterois* and *Brachirus* sp.) with much the same distribution have caused serious injuries to divers and aquarists. The California scorpionfish *(Scorpaena guttata)* is an abundant species, and 247 cases of stings were reported during the 1953-1960 period in southern California. Most of these were sustained by fishermen, "bait boys" who clean fish aboard sportsfishing boats, and housewives preparing fish for cooking (Russell, 1965). During the Sealab II project, scorpionfish became very numerous around the underwater shelter at a depth of 61 m, and several divers suffered stings (Halstead, 1970). A related species, *S. scropha,* is a frequent cause of stings in the Mediterranean (Maretic, 1957). The bullrout *(Notesthes robusta)* causes many injuries in the estuaries and coastal river systems of eastern Australia. Stonefish are a hazard around reefs in this area, and several other less dangerous species such as the rock-cods (*Ruboralga* sp.) and fortescues (*Centropogon* sp.) also occur.

Although all fish stings are painful, those of scorpaenids are *primus inter pares*. Steinitz (1959) wrote of his own experience with a lionfish sting, "I was tortured by pains beyond measure, and yet the pain was still growing more intense. . . . It is a strange experience recognizing quite lucidly that nothing fatal has happened . . . and feeling at the same time that this was much worse than anything previous. In fact it is just short of driving oneself completely mad." Smith (1951) speaking of a personal experience with stonefish sting says, "Before reaching the beach, only five minutes away, the pain was spreading through the hand . . . and was of an intensity never before experienced . . . there remains little recollection save of a grim battle to remain conscious and of an insane desire to ease the mounting agony by rolling on the ground." The pain persists for several hours and is accompanied by severe local swelling, sometimes with the formation of blisters. There may be necrosis and sloughing. Particularly with stonefish stings, systemic manifestations such as profuse sweating, dyspnea, hypotension, cyanosis, and collapse are characteristic. Vomiting and headache may follow the comparatively mild sting of the California scorpionfish. Deaths from stonefish sting have occurred within an hour (Smith, 1957) and usually take place within the first six hours. Those who survive a severe sting may complain of weakness, shortness of breath, and muscular aches for several weeks.

As with other fish stings, soaking the affected part in water as hot as can be tolerated is the most effective first aid measure. Ordinary analgesics and local anaesthetics seem to afford little relief from these stings, but prompt infiltration of the wound with emetine hydrochloride solution (65 mg per ml) is recommended (Phelps, 1960). A stonefish antivenin is produced by the Commonwealth Serum Laboratories of Australia, and a scorpionfish antivenin is available in limited quantity from the Medical Center of Pula, Yugoslavia. In the case of the Australian product, an intramuscular or intravenous dose of 2 to 4 ml is recommended. This will neutralize 32-64 mg of venom (Wiener, 1959). The similarity of clinical manifestations in scorpaenid stings suggests that stone-fish antivenin may be effective in treating severe envenomation by all fishes of this group. The antivenin is produced in horses, and the usual precautions against anaphylactic reactions should be observed.

B. Weeverfish

The weeverfish are a small group of venomous fishes whose distribution is confined to the eastern Atlantic and Mediterrane-an. Three of the species occur chiefly in shallow water with a sand or mud bottom; one, *Trachinus draco,* is primarily a deep water species. They are small fish with a maximum length of about 45 cm. Although they spend much time buried in mud or sand, they are more active than most venomous fish and tend to be aggressive (Halstead, 1957, 1970). They are food fish of com-mercial importance in some parts of Europe.

The venom apparatus is associated with the dorsal and opercu-lar spines and is essentially similar to that of the scorpaenids con-sisting of deeply grooved spines, elongate venom glands, and an integumentary sheath. The opercular glands are better developed than those of the dorsal spines.

Weeverfish venom free of contaminating material can be ob-tained by aspiration from the spine grooves of live or iced fish. It is a clear fluid with pH of 7.1. It rapidly loses its toxicity under ordinary conditions of storage but may be preserved in 15 per cent glycerine at −60° (Skeie, 1962a).

Weeverfish venoms contain three protein fractions, of which the one with the least electrophoretic mobility contains most of the toxin. In addition, cholinesterase, adrenalin, noradrenalin, and histamine have been reported in the crude secretion. Sero-tonin has been found in venom of the lesser weever (*Trachinus vipera*) but not in that of the greater weever (*T. draco*) (Skeie, 1962b; Haavaldsen and Fonnum, 1963).

Intravenous injection of lethal doses of weeverfish venom into mammals produces gasping respirations, a frothy discharge from the nose and mouth, a marked fall in blood pressure with bradycardia, electrocardiographic evidence of myocardial ischemia, and terminal convulsions. The lethal dose is difficult to establish because of the instability of venom preparations. Skeie (1966) reports the mouse LD_{50} of crude venom as 10 mcg or about 0.5 mg per kilo. Sublethal doses produce similar but milder effects on the cardiovascular and respiratory systems. Vasoconstriction is marked and may lead to loss of hair and gangrene of the ears and tail tip. Local necrosis is not uncommon (Russell and Emery, 1960; Skeie, 1962c).

Weeverfish stings, for the most part inflicted by *T. draco,* are fairly common among commercial fishermen in the North Sea and Mediterranean. Stings by *T. vipera* and the other small species are usually inflicted on bathers and skin divers. Spines of dead fish can cause injury for several hours especially if the fish is refrigerated.

Figure 27. The Greater Weeverfish *(Trachinus draco)* reaches a length of about 45 cm and is of some commercial importance in European waters. It has highly venomous dorsal and opercular spines. (Photo courtesy Z. Maretic.)

As with other fish stings, the principal manifestation of weeverfish envenomation is severe pain radiating centrally from the wound and lasting a few hours to several days. The spine punctures are surrounded by a zone of ischemia and edema. Systemic manifestations include sweating, fainting, disorientation, dyspnea, cyanosis, anginal pain, and cardiac arrhythmia (Skeie, 1966). Halstead (1957) reported the case of a diver stung on the jaw presumably by *T. vipera.* In addition to the above symptoms he developed a massive hematoma involving most of the head and

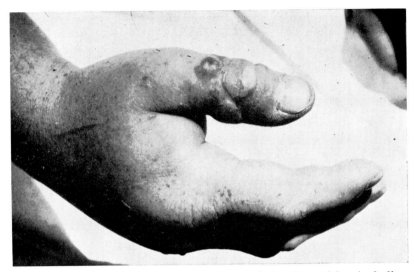

Figure 28. Weeverfish sting on dorsal aspect of thumb resulting in bullous lesions and edema. Photographed about 24 hours after injury. (Photo courtesy Z. Maretic.)

neck and upper chest. This was accompanied by severe respiratory embarrassment, fever, and leucocytosis. Deaths from weeverfish stings have been mentioned by several authors but without sufficient details to determine if they were due to toxemia or secondary infection.

Immersion of the injured area in hot water and intravenous injection of calcium gluconate have been reported as effective therapeutic measures; the general therapy is much like that of stingray envenomation. Weeverfish venom is antigenic but does not make a satisfactory toxoid. The production of an antivenin was reported to be impractical (Skeie, 1962b).

C. Venomous Catfish

The catfishes are a large group of more than a thousand species, most of them found in tropical and subtropical streams. A fair number of species are partially or exclusively marine. They frequently aggregate in dense schools. About fifty species have been reported as venomous.

In catfish, the venom-producing glandular cells ensheath the pectoral and dorsal spines which frequently are strongly serrate. There is an integumentary sheath. The amount of glandular tissue varies considerably among species. In general, the venom apparatus of the pectoral spines is the better developed. When the fish is disturbed, its dorsal and pectoral spines become locked into a rigid, extended position.

The chemistry of catfish venoms is virtually unknown, although there is some indication that they are more stable than other fish venoms. Venom gland extracts from the striped marine catfish *(Plotosus lineatus)* were reported to contain a dialyzable neurotoxic fraction and nondialyzable hemolytic fraction. Similar properties were reported for venom of *Heteropneustes fossalis,* an Indian freshwater catfish (Halstead, 1970).

Plotosus venom in experimental animals produces muscle spasm, respiratory distress, and local manifestations of erythema, cyanosis, and often necrosis. Rats, rabbits, sparrows and carp were relatively susceptible to the venom; loach, frogs and snakes moderately resistant; and newts highly resistant (Toyoshima cited by Halstead, 1970). Venoms of the freshwater catfish *Schilbeodes* and *Noturus* showed higher toxicity for mammals than for fish, while the reverse was true of *Ictalurus* venom (Birkhead, 1967).

Catfish envenomation of man usually occurs among fishermen removing the fish from hooks, nets, or traps. *Plotosus lineatus* with a wide distribution in the tropical and subtropical Indo-Pacific area can cause dangerous stings characterized by much pain and swelling, blisters, and some degree of shock. Herre (1949) reported severe diarrhea and a weight loss of 25 pounds following envenomation by this species, but intercurrent infection was not ruled out. *Cnidoglanis megastoma* is an Australian coastal and estuarine species that can inflict severe stings (Pacy, 1966). Stings by the small madtoms, *Schilbeodes* and *Noturus* sp., of the eastern and midwestern U. S. are painful but rarely require medical attention. *Galeichthys felis,* a marine and brackish water species of American coasts from Cape Cod to Panama, can produce a severe laceration accompanied by great pain. Members of the genera *Pimelodus* and *Clarias* have been reported to cause injuries in

Figure 29. The Brindle Madtom *(Noturus miurus)* is a small venomous catfish native to streams in the midwestern United States. Its sting is painful but not dangerous. (Illustration courtesy Illinois Natural History Survey.)

Brazil (Halstead et al., 1953; Sawaya, 1966). Wounds caused by catfish often become secondarily infected with potentially serious consequences.

The treatment of catfish stings is symptomatic. General measures as advocated for stingray envenomation are helpful. Prophylactic measures against tetanus and other bacterial infections are important.

D. Toadfishes

The toadfishes are a widely distributed group of small, bottom-dwelling fishes with a superficial resemblance to scorpionfish and stonefish. Most are marine or estuarine, but a few occur in rivers. They occur in turbid, shallow water and are sedentary in habits.

The venom apparatus of toadfishes consists of dorsal and opercular spines surrounded by venom producing glandular tissue. The spines are unique among venomous fishes in being hollow with a large opening at the base and a smaller one at the tip. There are no ducts, but venom squeezed from the gland when the spine penetrates tissue is injected through the distal opening into the wound.

Venom of the Brazilian toadfish, *Thalassophryne* sp., was described as slightly acid, clear, and albuminous. Injected into guinea pigs, it produced local necrosis, mydriasis, ascites, and paralysis of the hind limbs (Froes, 1933) .

Toadfish stings are infrequent and apparently present no distinctive clinical features being in most respects similar to scorpion-fish and catfish stings. Serious envenomation has not been reported.

E. Miscellaneous Venomous Fishes

Venomous species have been reported in more than twenty additional families of fishes, although in many instances the evidence is not very convincing. The stargazers (Uranoscopidae) have venomous shoulder spines, while rabbitfish (Siganidae) have no less than 24 grooved spines with venom-secreting tissue. Surgeonfish (Acanthuridae) can inflict painful wounds with their caudal spines, although evidence for an associated venom is not

Figure 30. One of the venomous Rabbitfish *(Lo vulpinus)*. Fishes of this group can inflict stings with their dorsal, pelvic and anal spines. (Photo by author.)

altogether convincing. Of considerable biological interest is the report that the leatherback or lae *(Scomberoides sanctipetri)* may use its venomous anal spines to incapacitate or kill fish on which it preys. It is a fast swimming, midwater fish whose habits are quite different from those of most venomous fishes (Halstead et al., 1972) .

The differential diagnosis of injuries by spines of venomous fishes may be difficult if the fish was not seen or captured. Coelenterate stings do not show puncture wounds, and the lesions are frequently linear. Injuries by spines of sea urchins do not produce the intense, centrally radiating pain of fish stings. Sea snake bites are almost painless, and the onset of systemic symptoms is usually delayed 30 minutes or more. Stingray wounds are usually larger and more ragged than those inflicted by the venomous bony fishes. One or more puncture wounds accompanied by prompt onset of intense pain, sweating, dyspnea, and collapse suggest envenomation by a stonefish or one of the other more dangerous scorpaenids.

Chapter 5

SNAKES AND SNAKE VENOMS

INTRODUCTION

S NAKES ARE A DISTINCTIVE and specialized group of reptiles that diverged from ancestral lizards near the end of the Cretaceous and now are represented by about 2700 species. Evidence indicates the primitive snakes were nonvenomous; the oldest venomous snake fossils have been found in the Miocene.

Venomous snakes are found throughout most of the world, although they barely enter the arctic in Scandinavia and are completely absent from the antarctic and a number of major islands or island groups, e.g. New Zealand, Malagasy, most of the West Indies. Marine species are confined to the Indian and Pacific Oceans, mostly in coastal waters.

The biology of venomous snakes is in no way different from that of snakes in general. They may be arboreal, terrestrial, aquatic, or burrowing. All are secretive, and a majority of tropical species are nocturnal. Temperate zone species hibernate in cold weather, sometimes assembling in large numbers at this time. The venomous snakes are almost equally divided between egg-laying and live-bearing species. A few of the oviparous species give some degree of parental care to their eggs. All snakes are carnivores and feed very largely on live animals, although some species scavenge occasionally. Venom in snakes almost certainly evolved as an adaptation for subduing prey. This remains its primary role in nearly all species; it serves secondarily as a defensive adaptation.

While there are no rules of thumb for identification of poisonous snakes on a world wide basis, a "poisonous vs. harmless" identification can often be made by any biologically knowledgeable individual. In a previous publication, I have summarized

some of the principles (Minton, 1970). Handbooks for identification of the poisonous snakes of most geographic regions are available from natural history museums or government agencies. Local
information about reptiles is often untrustworthy; erroneous
beliefs are widespread even among educated and professional persons. Harmless reptiles are often believed dangerous, while
dangerous species may occasionally be considered innocuous.

The major taxonomic subdivisions of snakes are listed in
Table IV. I have departed from the widespread practice of considering the pit vipers a family distinct from the Old World vipers
that lack the heat-sensing pits. This can be justified on the
grounds of purely morphological taxonomy; moreover, there are
no significant biochemical differences between the venoms of the
two groups, nor are there distinctive clinical pictures of envenomation that characterize them. Differences do exist but are more
evident at the genus or species level. Snakes with aberrant venoms,
e.g. Wagler's pit viper *(Trimeresurus wagleri)* and the berg adder
(Bitis atropos) occur in both groups. While the mole vipers have
viperid dentition, they are highly aberrant in other ways and de-
serve recognition as a separate group. They are of minor clinical
importance. Despite their relationship to the elapids (cobras and
their kin), I consider the sea snakes a distinct family. They are
highly specialized morphologically and physiologically, and their
venoms differ biochemically from those of land snakes. Sea snake
envenomation clinically and epidemiologically is a distinct entity.

The venom apparatus of snakes consists of grooved or tubular
enlarged maxillary teeth and associated venom glands. The venom
glands evidently evolved from oral glands whose original function
was to lubricate food for ease in swallowing and to secrete digestive enzymes. The importance of venom in digestion is unknown.
A very limited amount of data indicates that poisonous snakes
whose venom ducts have been ligated digest food normally.

It would appear that the ophidian venom apparatus has
evolved independently at least twice. In the Colubridae, the enlarged, grooved fangs are at the posterior end of the maxillary
bone, if present. The venom or Duvernoy's gland is in the posterior supralabial region. This gland was found in 102 of 118

TABLE IV. MAJOR GROUPS OF SNAKES

Group	Distribution	Remarks
Cylinder or pipe snakes Family Aniliidae	Southeast Asia, tropical America	Probably the most primitive of snakes. Medium-size burrowers. Not venomous
Boas and pythons Family Boidae	Mostly in tropical and warm temperate zones. Pythons in Old World only	Includes both large and small species. Not venomous
Shield-tail snakes Family Uropeltidae	South India and Ceylon	Small, burrowing snakes. Not venomous
Sunbeam snake Family Xenopeltidae	Southeast Asia	Single, primitive species. Medium-size; not venomous
File snakes Family Acrochordidae	Southeast Asia to northern Australia in estuarine and coastal habitats	Two strictly aquatic species. Not venomous but may be confused with sea snakes.
Blind snakes Families Typhlopidae and Leptotyphlopidae	Tropical and warm temperate zones	Very small, worm-like snakes. None venomous
"Typical" snakes Family Colubridae	Almost worldwide except for Arctic, Antarctic, south Australia, and certain islands	Large and extremely varied family. Many species with venom glands and posterior maxillary fangs, but very few capable of causing clinically significant envenomation.
Cobras, mambas, coral snakes, kraits, etc. Family Elapidae	Tropical and warm temperate zones	About 180 species. All venomous; fangs at anterior end of maxillae
Sea snakes Family Hydrophiidae	Mostly southeast Asian and Australian coastal waters	About 50 species; nearly all marine or estuarine. All venomous; fangs similar to those of Elapidae; venoms highly toxic
Vipers Family Viperidae	Almost worldwide except for Australian region, Antarctic, and certain islands	Three major subdivisions. All venomous; fangs on much reduced maxillae and highly movable
Pit vipers Subfamily Crotalinae	The Americas and much of Asia, extreme eastern Europe	About 120 species. Heat sensing pits between eyes and nostrils
Old World vipers Subfamily Viperinae	Africa, Europe, and Asia	About 40 species. No specialized heat sensing pits
Mole vipers Subfamily Atractaspinae	Africa, limited areas in Middle East	About 15 species. Fangs often used singly with backward, stabbing motion. Venoms significantly different from those of other vipers

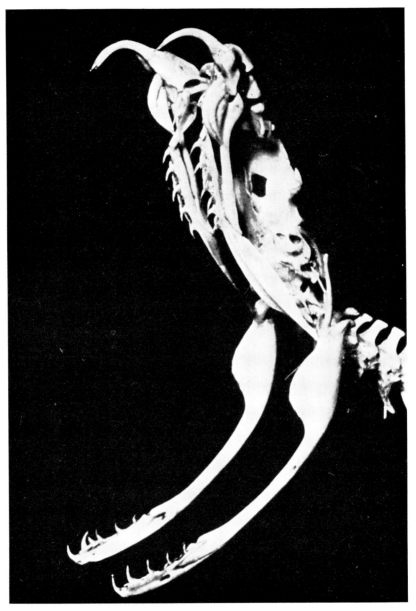

Figure 31. Rattlesnake skull with fangs swung foreward as in striking. Note solid teeth on palatine and dentary bones. (Photo by John Tashjian.)

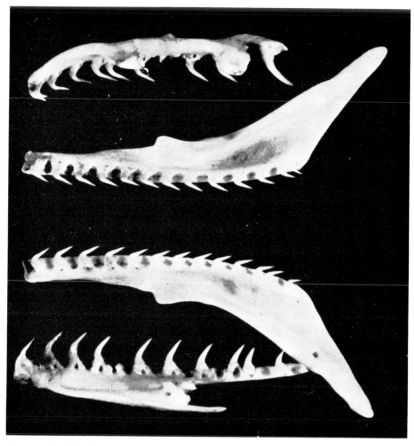

Figure 32. Tooth-bearing bones and dentition of a typical nonvenomous *snake (Natrix rhombifera.)* Top: right maxilla. Center: right and left palatines. Bottom: dentary. The most posterior maxillary tooth is slightly enlarged in this species, but no groove for conducting venom is present. (×3) (Photo by Illustration Department, Indiana University Medical Center.)

colubrid genera and showed great variability in gross and microscopic anatomy. It may or may not be associated with enlarged, grooved fangs (Taub, 1967). The type of venom apparatus possessed by most snakes whose bite is dangerous to man consists of grooved or tubular fangs at the anterior end of the maxillary bone. In the Viperidae, the maxillary is greatly reduced and bears

Figure 33. Left maxilla of Taipan *(Oxyuranus scutellatus),* a large Australian elapid snake. Note large fang, short maxilla, and posterior small tooth. A second posterior tooth, normally present, is missing in this specimen. (×10) (Photo by Illustration Department, Indiana University Medical Center.)

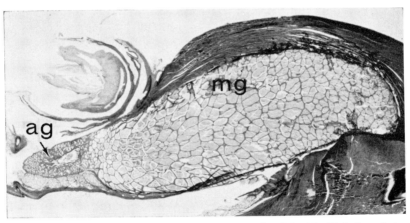

Figure 34. Longitudinal section of cobra venom gland (×16). Tubules of the main venom gland (mg) empty into a duct that leads to the fang. The accessory gland (ag) surrounds this duct. The muscle on the dorsal surface of the gland is involved in expelling the venom. (Preparation and photo courtesy H. I. Rosenberg.)

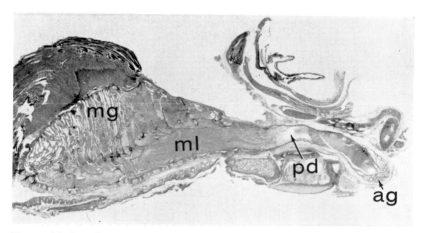

Figure 35. Longitudinal section of rattlesnake venom gland (×18). Tubules of the main gland (mg) empty into a reservoir (ml) continuous with the duct (pd) leading to the fang. The accessory gland (ag) surrounds the duct just before it enters the fang. Photo courtesy H. I. Rosenberg from a preparation by E. Kochva.)

no other teeth, and the fang has a wide range of movement. In the Elapidae and Hydrophiidae, the maxillary is slightly to moderately reduced in length, frequently bears other teeth, and the fang has less mobility than in the Viperidae. The venom glands in these three families also occupy the posterior supralabial region except in a few species where they are greatly elongate and tubular extending far behind the head. The venom glands of the Elapidae, Hydrophiidae, and Viperidae appear to have a different origin than Duvernoy's gland (Taub, 1966). It cannot be determined at present if the viperid and elapid venom apparatus had a common origin or developed independently; the hydrophid venom apparatus is essentially elapid in character. A special muscle, the compressor glandulae, facilitates emptying of the venom gland in most Viperidae; the superficialis muscle has this function in the Elapidae.

The main viperid venom gland is roughly triangular in shape and formed of repeatedly branching tubules arranged around a central lumen in which considerable quantities of venom can be stored. A duct extends anteriorly to the base of the fang, and a small accessory gland surrounds the duct just before it opens into the fang. Several types of cells are found in the main gland and two types in the accessory gland. There seem to be no significant differences between the venom glands of pit vipers and other vipers. Details of the histology and histochemistry of these glands have been described by Kochva and Gans (1965, 1966, 1970), Gennaro et al. (1960), and Schaeffer et al. (1972a, 1972b). Secretory activity is greatest just after the gland has been emptied and virtually ceases when the gland is full. Numerous enzymes have been identified in the secretory cells of the main gland; mucins are secreted by cells in the anterior portion of the accessory gland. Several types of secretion vesicles and free ribosomes have been observed in pit viper venom within the gland lumen (Gennaro et al., 1968).

Elapid and sea snake venom glands have parallel tubules lined with secretory epithelium in the main part of the gland. The duct draining the main gland passes through the center of the mucus-secreting accessory gland which is directly anterior to it. There is less space in the gland for venom storage. The process of venom secretion seems to be essentially the same as in viperids (Rosenberg, 1967). Additional histochemical information chiefly based on study of glands of the desert blacksnake *(Walterinnesia)* is provided by Kochva and Gans (1970).

Venom glands of the atractaspids or mole vipers have a central lumen surrounded by essentially unbranched tubules. There is no accessory gland, but mucus-secreting cells are present at the luminal ends of the secretory tubules. The secretory cells are extraordinarily fragile and unique in some of their histochemical reactions (Kochva et al., 1967).

Venom may be extracted with comparatively little difficulty from viperids and larger elapids; it is more difficult to obtain from small elapids, sea snakes, and colubrids. Description of techniques have been given by Bücherl (1963), Deoras (1963), Ashley and Burchfield (1968), and Gans and Elliott (1968). Snake venoms are colorless to dark amber liquids having a viscosity of 1.5 to 2.5 and specific gravity of 1.03 to 1.12. For most species the pH ranges from 5.5 to 7.0. The percentage of solids varies from 18 to 67.

Venom yield is highly variable and influenced by many factors such as species of snake, size, age, season, and general health of the animal. Yields for representative species are given in Table V.

BIOCHEMISTRY AND PHARMACOLOGY
Enzymes

No biological toxins equal snake venoms in biochemical and pharmacological complexity. It is hard to avoid the hypothesis that in the biochemical evolution of venom, snakes have added new compounds while retaining others that may represent an earlier phylogenetic stage of development. The ubiquity of many of these substances in snake venoms suggests that they must have adaptive significance; however, this is by no means evident in all cases. The pharmacologically active substances in snake venoms can be divided into the following categories: (1) enzymes (2) polypeptide toxins (3) glycoproteins (4) low molecular weight compounds. It should be emphasized that the best known snake venoms are those of a comparatively small number of common and medically important vipers, pit vipers, and elapids. Less is known of sea snake venoms and very little of colubrid and mole viper venoms.

An interesting feature of snake venoms is the high stability of many of their enzymes and toxins when stored in a dry state. Rattlesnake venom stored 50 years had approximately the same LD_{50} as freshly extracted venom (Russell and Eventov, 1964). Other snake venoms have retained lethality and some enzyme activities during 25 to 39 years storage, sometimes at room temperature (Smith and Hindle, 1931; Sugihara et al., 1972). Most properties of snake venom are retained well in promptly lyophilized material.

At least 17 enzymes, most of them hydrolases, have been detected in snake venoms. Ten have been found in all venoms that have been adequately investigated; the others occur chiefly in certain taxonomic groups or are characteristic of only a few species. Since immunological studies of venoms generally show an absence of common antigens in venoms of remotely related species such as cobras and rattlesnakes, the widely distributed venom enzymes

TABLE V. VENOM YIELD AND TOXICITY OF SOME IMPORTANT
VENOMOUS SNAKES

Species	Average adult length cm	Venom yield mg	Mouse LD$_{50}$ I-V	Mouse LD$_{50}$ I-P	mg/kilo S-C
Boomslang *Dispholidus typus*	120-150	4-8	0.10		12.5
Northern mole viper *Atractaspis microlepidota*	50-60	5-10			4.5
Death adder *Acanthophis antarcticus*	50-75	70-100	0.25		0.50
Australian brown snake *Pseudonaja textilis*	150-175	5-10	0.01		0.25
Australian copperhead *Denisonia superba*	90-140	10-35	0.30		1.12
Taipan *Oxyuranus scutellatus*	170-230	100-200	0.02		0.12
Tiger snake *Notechis scutatus*	90-150	30-70	0.03	0.04	0.15
Australian blacksnake *Pseudechis porphyriacus*	125-175	30-50	0.50		2.0
Mulga snake *Pseudechis australis*	150-200	150-300	0.30		1.5
Indian krait *Bungarus caeruleus*	90-120	8-12	0.09	0.09	0.45
Many-banded krait *Bungarus multicinctus*	90-120		0.07	0.08	0.19
Banded krait *Bungarus fasciatus*	125-160	25-50	1.38	1.55	3.6
Eastern green mamba *Dendroaspis angusticeps*	175-225	60-95	1.50		3.05
Black mamba *Dendroaspis polylepis*	180-250		0.25		0.32
Indian cobra *Naja naja naja*	120-155	170-250	0.24	0.28	0.29
Chinese cobra *Naja naja atra*	115-150	150-200	0.22		0.28
Philippine cobra *Naja n. philippinensis*	100-140				0.14
Cape cobra *Naja nivea*	145-175	100-150	0.35		0.40
Egyptian cobra *Naja haje*	150-200	175-300	0.95		1.75
Spitting cobra *Naja nigricollis*	150-200	200-350	1.15		3.05
Ringhals *Hemachatus haemachatus*	100-130	80-120	1.10		2.65

TABLE V.   VENOM YIELD AND TOXICITY OF SOME IMPORTANT
VENOMOUS SNAKES

Species	Average adult length cm	Venom yield mg	Mouse LD$_{50}$ I-V	I-P	mg/kilo S-C
King cobra	210-360	350-500	1.60	1.64	1.73
Ophiophagus hannah					
North American coral snake	58-82	3-5	0.38	0.67	1.30
Micrurus fulvius					
Brazilian giant coral snake	100-120	20-30			2.50
Micrurus frontalis					
Beaked sea snake	90-125	10-15	0.12	0.11	0.15
Enhydrina schistosa					
Annulated sea snake	135-165	5-9	0.35	0.24	0.65
Hydrophis cyanocinctus					
Hardwicke's sea snake	75-100	2-4	0.20	0.27	
Lapemis hardwicki					
Rhombic night adder	56-72	20-30	9.25		15.0
Causus rhombeatus					
Puff adder	90-125	160-200	0.35	3.65	7.75
Bitis arietans					
Gaboon viper	125-155	450-600	0.70	1.96	12.5
Bitis gabonica					
Desert horned viper	52-65	40-70	0.50		15.0
Cerastes cerastes					
Saw-scaled viper	40-56	20-35	2.30		6.55
Echis carinatus					
Russell's viper	100-125	130-250	0.08	0.40	4.75
Vipera russelii					
Levantine viper	80-115	75-150			16.0
Vipera lebetina					
Palestine viper	90-120	90-140	0.18		9.40
Vipera xanthina palaestinae					
Long-nosed viper	62-76		0.45		6.59
Vipera ammodytes					
European viper	48-62	10-18	0.55	0.80	6.45
Vipera berus					
Malay pit viper	58-82	40-60	4.71	4.98	23.35
Agkistrodon rhodostoma					
Hundred-pace snake	90-120				17.25
Agkistrodon acutus					
Mamushi	50-66			1.32	20.00
Agkistrodon halys blomhoffi					
Cottonmouth	76-115	100-150	4.00	5.11	25.80
Agkistrodon piscivorus					
Copperhead	60-90	40-70	10.92	10.50	25.60
Agikistrodon contortrix					

TABLE V. VENOM YIELD AND TOXICITY OF SOME IMPORTANT
VENOMOUS SNAKES (Continued)

Species	Average adult length cm	Venom yield mg	Mouse LD$_{50}$ I-V	I-P	mg/kilo S-C
Terciopelo, Fer-de-lance	115-175	150-300	4.27	3.80	22.00
Bothrops atrox					
Urutu	90-115	60-100	1.96		15.80
Bothrops alternatus					
Jararaca	110-140	40-70	1.12		7.00
Bothrops jararaca					
Jararaca pintada	65-110	25-40	2.31		14.00
Bothrops neuwiedi					
Jararacussu	110-165	100-200	0.46		13.00
Bothrops jararacussu					
Eyelash viper	45-64	10-20	1.98	3.63	33.15
Bothrops schlegeli					
White-lipped tree viper	40-64	8-15	3.75		12.75
Trimeresurus albolabris					
Wagler's pit viper	76-90	65-90	0.75	3.57	4.63
Trimeresurus wagleri					
Okinawa habu	125-180	100-300	4.25	5.07	27.30
Trimeresurus flavoviridis					
Chinese habu	80-110		0.64		
Trimeresurus mucrosquamatus					
Bushmaster	150-220	200-300	4.51	6.41	36.90
Lachesis mutus					
Eastern diamondback rattlesnake	110-165	250-500	1.68	1.89	14.55
Crotalus adamanteus					
Western diamondback rattlesnake	95-160	200-300	4.20	3.71	13.45
Crotalus atrox					
Timber rattlesnake	90-140	100-200	2.63	2.91	9.15
Crotalus horridus					
Prairie rattlesnake	90-115	50-100	1.61	2.25	16.15
Crotalus viridis viridis					
Pacific rattlesnake	80-125	75-160	1.29	1.60	3.56
Crotalus viridis helleri					
Mexican west-coast rattlesnake	120-150	250-350			2.78
Crotalus basiliscus					
Brazilian rattlesnake	100-125	25-40	0.13	0.30	0.60
Crotalus durissus terrificus					
Mojave rattlesnake	75-100	50-90	0.21	0.23	0.31
Crotalus scutulatus					
Pigmy rattlesnake	40-56	20-30	2.80	6.84	24.25
Sistrurus miliarius					

Venom yields are those reported for adult snakes of average size. Maximum yields may be two to five times the upper limit of the average. Venom toxicity is subject to considerable individual and geographic variation. See text.

evidently differ in molecular structure between these taxa. However, antigenically identical enzymes may be shared by related species, for example the esterase hydrolyzing carbonaphthoxycholine is the same in four species of Asian and African cobras, the king cobra, two species of kraits, and the North American coral snake, all of which are elapids (Munjal and Elliott, 1972).

Phospholipase A (Phosphatide acyl-hydrolase) is one of the most widespread of snake venom enzymes. It has been detected in a great variety of species including colubrids such as the boomslang *(Dispholidus)* and the cat-eyed snakes *(Leptodeira)*. It is present in exceptionally high concentration in venoms of *Naja nigricollis, N. melanoleuca,* and *Micrurus fulvius* which are elapids and *Bitis cornuta* and *Agkistrodon piscivorus* which are viperids. An interesting situation exists with the mambas *(Dendroaspis)* where the enzyme is present in high concentrations in venoms of *D. polylepis* and *D. viridis* but is virtually absent in *D. angusticeps* and *D. jamesoni* (Christensen 1955, 1968; Schwick and Dickgiesser, 1963; Kocholaty et al., 1971). Although single, homogenous enzyme preparations have been isolated from some venoms such as that of the sea snake *Laticauda semifasciata* (Tu and Passey, 1972), the activity often resides in a group of isoenzymes separable by electrophoresis or ion-exchange chromatography but otherwise indistinguishable. Seven such isoenzymes have been reported in *Naja naja* venom (Braganca and Sambray, 1967), and phospholipase A activity was present in all the seven fractions obtained from *Micrurus fulvius* venom (Ramsey et al., 1972). Snake venom phospholipase A enzymes are highly heat stable and resistant to tryptic hydrolysis, properties that may be related to the high degree of cross-linking in the molecule. In some venoms such as that of the tropical rattlesnake, *Crotalus durissus,* phospholipase A forms a complex with a polypeptide toxin. Several phospholipase A preparations from snake venoms appear to be homogenous and have been well characterized (Table VI).

Multiple biological activities have been attributed to snake venom phospholipase A. Through its hydrolysis of lipoprotein-bound phospholipids, it inhibits mitochondrial electron transfer systems, uncouples oxidative phosphorylation, and produces phos-

Venom Diseases

TABLE VI. ENZYMES AND SOME OTHER BIOLOGICALLY
ACTIVE SUBSTANCES ISOLATED FROM
SNAKE VENOMS

Substance	Snake Species	Remarks	Reference
Phospholipase A	*Agkistrodon halys blomhoffi*	Partial amino acid sequence determined	Samejima et al 1971
"	*Naja nigricollis*	Two enzymes; M.W.'s 13,000 & 14,600	Wahlstrom 1971
"	*Crotalus adamanteus*	M.W. ca 30,000; 261 amino acid residues	Wells & Hanahan 1969
"	*Crotalus atrox*	M.W. 14,500. Requires divalent metal ions	Wu & Tinker 1969
"	*Bothrops neuweidii*	Two isoenzymes similar to those in other *Bothrops* sp	Vidal & Stoppani, 1971
"	*Laticauda semifasciata*	M.W. 11,000	Tu & Eassey 1972
Protease A	*Bitis arietans*	M.W. 21,400. Requires Ca^{++}	Van Der Walt & Joubert 1971
Procoagulant	*Crotalus admanteus*	Acts directly on fibrinogen	Damus et al 1972
"	*Echis carinatus*	Converts prothrombin to thrombin, M.W. 86,000	Schieck et al 1972
Leucostoma peptidase A	*Agkistrodon piscivorus leucostoma*	M.W. 22,500. Amino acid composition determined	Wagner et al 1968
Hemorrhagic Principle I	*Trimeresurus flavoviridis*	M.W. ca 100,000 Mouse LD$_{50}$ 4.63 mcg	Omori-Satoh & Ohsaka 1970
Anticoagulant Principle	*Agikistrodon acutus*	Glycoprotein M.W. 20,650; 160 amino acid residues	Ouyang & Teng, 1972
Complement Factor (CoF)	*Naja naja*	Glycoprotein M.W. 144,000	Müller-Eberhard & Fjellström 1971

phate acceptor inhibition. Venoms of various species vary markedly in this activity (Ziegler et al., 1967). Phospholipase A has long been associated with the hemolytic activity of snake venoms and is the "indirect hemolysin" of many investigators. Its hemolytic activity *in vitro* depends upon its interaction with other factors (Condrea and de Vries, 1965). There is no proof that phospholipase A is involved in the intravascular hemolysis that accompanies envenomation by certain snakes; however, the phospholipase A of *Vipera xanthina palestinae* venom produced sphering and other abnormalities of erythrocytes in sublethal doses (de Vries et al.,

1962). These abnormalities have been observed in clinical cases of envenomation (Perkash and Sarup, 1972; Weiss et al., 1973). It has been suggested that the phospholipase A of *Enhydrina schistosa* venom which has little affinity for erythrocytes may attack membrane phospholipids of muscle cells and account for the myolytic activity of this sea snake venom (Ibrahim, 1970). It is also believed to participate in the local hemorrhage, edema, and myolysis caused by the venom of the pit viper *Trimeresurus flavoviridis* (Maeno et al., 1960, 1962).

Venom phospholipase A is known to cause mast-cell degranulation and disintegration of platelets and leukocytes, all of which are associated with the release of histamine (Hogberg and Uvnas, 1960). This is seen with a great number of snake venoms, although it is not known if it is always mediated through the same mechanism. The phospholipase A of certain cobra venoms is reported to inactivate the Christmas factor which is involved in normal blood coagulation (O'Brien, 1956).

The role of phospholipase A in the neurotoxic effects of snake venoms is controversial. Most of the highly purified preparations of the enzyme prepared in recent years have been completely without neurotoxic effects, indeed they have had very little toxicity of any sort. They may, however, facilitate the penetration of venom neurotoxins into nerve tissue.

The effects of venom phospholipase A on membranes extend to the lipid-containing envelopes of certain viruses. Inactivation of Murry Valley encephalitis virus by *Pseudechis porphyriacus* venom (Anderson and Ada, 1960) and of Friend leukemia virus by *Naja naja* venom (Okabe et al., 1964) has been reported and ascribed to this enzyme. Animals infected with Rauscher leukemia virus may be protected by cobra venom or a nontoxic fraction of the venom with phospholipase A activity (Raitano, 1968). The effects of *Naja* and *Bungarus* venoms on the ultrastructure of this virus have been reported by Padgett and Levine (1966).

Phospholipase B (Lysolecithin acyl-hydrolase) was first reported in venoms of certain Australian snakes, notably *Pseudechis porphyriacus, P. australis,* and *Denisonia superba,* as well as in venoms of some species from other parts of the world (Doery and

Pearson, 1964). It was reported in venoms of seven species of
Egyptian snakes (Mohamed et al., 1969b) and may be quite
widely distributed. Like phospholipase A, it is thermostable and
requires divalent metallic ions. Its maximal activity is at pH 8.0
to 10.0 in contrast to 7.0 to 9.0 for most snake venom phospholipase A preparations. Phospholipase C (Phosphatidylcholine cho-
linephosphohydrolase) activity has been reported in Asian cobra
venom (Braganca and Khandeparkar, 1966) and in the venom of
the South American pit viper *Bothrops alternatus* (Sarkar and
Devi, 1968). Virtually nothing is known of the pharmacological
activity of these snake venom phospholipases nor have they been
adequately isolated and characterized.

Proteolytic enzymes are present in high concentration in
venoms of most vipers and are believed important in the pathogenesis of viperid envenomation. With certain exceptions, they
are absent or present in only small amounts in venoms of sea
snakes and elapids. Jimenez-Porras (1970) has divided these
enzymes into heat-labile, acidic proteases; heat-stable, nonproteo-
lytic, amino acid-ester hydrolases; and dipeptide and tripeptide
hydrolases. Viper venom proteases are similar to trypsin in molecular weight and in showing optimum activity at alkaline pH.
However, they are not affected by such trypsin antagonists as
ovomucoid and soybean trypsin inhibitor but are inhibited by
ethylenediamine-tetracetate (EDTA). Removal of divalent metallic ions, zinc and calcium chiefly but copper, iron and manganese
in some venoms, greatly reduced proteolytic activity (Tu et al.,
1966; Friederich and Tu, 1971). Addition of divalent metallic
ions did not increase the protease activity of *Crotalus* and *Agkistrodon* venoms; some ions such as mercury inhibited activity
(Wagner and Prescott, 1966). These proteases are probably non-
serine-type metalloproteins. As determined by casein hydrolysis,
endopeptidase activity was particularly high in certain rattlesnake venoms such as *Crotalus atrox* and *C. basiliscus,* appreciably
lower in venoms of Old World vipers, and very low in elapid
venoms except for that of the king cobra, *Ophiophagus* (Oshima
et al., 1969). The instability of snake venom proteases has hin-
dered their isolation and characterization; however, several puri-

fied and at least partly characterized preparations have been reported (Table VI) .

The amino acid esterases of snake venoms are nonmetalloprotein serine enzymes (Wagner et al., 1968) that hydrolyze such synthetic substrates as benzoyl-L-arginine methylester (BAME) and p-toluenesulfonyl-L-arginine methyl ester (TAME) but not L-arginine methyl ester (AME) which is the most specific trypsin substrate. Arginine esterase activity has been observed almost exclusively in viperid venoms; however, elapid venoms can hydrolyze other dipeptide substrates (Tu et al., 1967; Tu and Toom, 1967a, 1968) . Exopeptidase activity as evidenced by ability to hydrolyse L-leucyl-B-napthylamide was found to at least a slight degree in 49 venoms from vipers, pit vipers, elapids, and sea snakes. It was particularly high in venoms of *Dendroaspis angusticeps, Oxyuranus scutellatus,* and *Echis carinatus* (Tu and Toom, 1967b) .

Proteolytic enzymes have generally been given a major role in pathogenesis of the hemorrhagic, necrotizing, and hypotensive effects produced by many snake venoms. It is becoming increasingly apparent, however, that these activities are seldom due to single enzymes but result from the interaction of several substances. Moreover, the effects of a crude venom may not reflect those of its component toxins. An extremely potent protein hemorrhagin (minimal hemorrhagic dose 0.0058 mcg.) has recently been isolated from the venom of *Trimeresurus flavoviridis* (Omori-Satoh and Ohsaka, 1970) , yet the hemorrhagic effect of crude venom is considerably less than that of several other venoms, and its most striking effect in the intact animal is myonecrosis (Homma and Tu, 1971) . The principal protease of *T. flavoviridis* venom is completely without hemorrhagic activity (Takahashi and Ohsaka, 1970) .

The action of viperid venom hemorrhagins appears to be mainly on the intercellular cement substance of the smaller blood vessels, for hemorrhage seems to occur through a morphologically intact vascular wall. With small doses of venom injected subcutaneously or intramuscularly, the extravasation of blood is confined to the site of injection; with large doses or with intravenous injection, the lung is the principal target organ (Jimenez-Porras, 1968). With some venoms such as that of *Agkistrodon halys blomhoffi,* there is a close correlation between hemorrhagic and lethal activity, but this is not the case with the majority of venoms. Venoms with particularly high hemorrhagic activity in experimental animals are those of *Bitis arietans, B. gabonica, Crotalus adamanteus, Agkistrodon halys, Bothrops picadoi,* and *Trimeresurus erythrurus* (Ohsaka et al., 1966; Homma and Tu, 1971). All these snakes are vipers or pit vipers. The colubrid, *Dispholidus typus,* has a strongly hemorrhagic venom whose mechanism of action is not well understood, although it hydrolyzes casein and TAME (Robertson and Delpierre, 1969). Venoms of some elapids such as *Pseudechis porphyriacus* also display hemorrhagic activity. *Walterinnesia* venom has little local hemorrhagic effect but produces pulmonary hemorrhage (Gitter et al., 1962). Considerable variation may occur between closely related snakes. Venoms of *Vipera russelii siamensis, Pseudocerastes persicus fieldi,* and *Crotalus durissus terrificus* have little or no hemorrhagic activity, while those of their subspecies *Vipera r. russelii, Pseudocerastes p. persicus,* and *Crotalus durissus totonacus* are moderately to highly hemorrhagic. In all these instances, the less hemorrhagic venom is the more toxic.

Interference with the normal mechanism of blood coagulation is a property of many snake venoms. The venom may display several types of action either sequentially or simultaneously. The coagulant action may be due to: (1) conversion of Factor X to activated Factor X (2) conversion of prothrombin to thrombin in the presence of Factor V (3) conversion of prothrombin to thrombin without added Factor V (4) conversion of fibrinogen to fibrin. Venoms of *Bothrops atrox, B. jararaca,* and *Agkistrodon rhodostoma* have both a Factor X activator and a thrombin-like activity converting fibrinogen to fibrin. However *Vipera russelii* venom contains only a Factor X activator, while *Crotalus durissus terrificus* venom has only thrombin-like activity. Venoms of *Oxyuranus scutellatus* and *Demansia (=Pseudonaja) textilis* have neither activity but convert prothrombin to thrombin in the absence of Factor V. Venoms of *Notechis scutatus* and *Acanthophis antarcticus* convert prothrombin to thrombin only in the presence of Factor V. (Denson, 1969). Anticoagulant properties of venoms are geenerally attributed to proteolytic enzymes that lyse either fibrin or fibrinogen or to the action of phospholipase on platelet or plasma phospholipid. A thrombin-like activity causing hyper-coagulability followed by incoagulability due to destruction of fibrinogen is seen with venoms of *Echis carinatus* and *Cerastes cerastes* among others (Mohamed et al., 1969a, 1969c). The anti-coagulant action of *Naja nigricollis* venom has been attributed to inhibition of thromboplastin plus heparin release. The activity is heat labile and apparently not due to phospholipase A (Mohamed et al., 1971). An enzyme with both thrombin-like and fibrinogen-degrading activity has been isolated from venom of *Crotalus ada-manteus* (Damus et al., 1972). The thrombin-like coagulant fac-tors of *Bothrops jararaca* and *Agkistrodon rhodostoma* venoms have been identified as esterases. The coagulant of *Vipera russelii* venom is an arginine esterase with a molecular weight over 100,000 (Meaume, 1966; Esnouf and Tunnah, 1967; Williams and Esnouf, 1962). The venom of *Agkistrodon acutus* owes its anticoagulant activity to a glycoprotein without protease or phospholipase activ-ity. Its mechanism of action is unknown, but it is potentiated by the fibrinolysin of *A. acutus* venom (Ouyang and Teng, 1972). *Dispholidus typus* venom is strongly coagulant and causes afibrino-genemia. The coagulant factor is believed to be an arginine esterase. The venom of this colubrid is surprisingly viperid-like in its hemorrhagic and anticoagulant activity (Robertson and Del-pierre, 1969; Mackay et al., 1969).

The coagulant action of snake venoms may cause rapid death by intravascular coagulation when large doses are given intra-venously, but concomitant fibrinolytic activity usually prevents this and leads to the production of a defibrination syndrome with non-clotting blood (Reid, 1967). If the vascular damage is mild and local as with *Agkistrodon rhodostoma* venom, the coagulation defect may have little deleterious effect, but if combined with a strong vasculotoxic agent as in *Echis* venom, multiple hemor-rhages into viscera and skeletal muscle are seen.

Bradykinin, a peptide with powerful hypotensive and smooth

muscle stimulating properties, is released from plasma by many viperid venoms and a few elapid venoms. Venoms particularly high in this activity are those of *Agkistrodon contortrix, Crotalus atrox* and *Lachesis mutus* (Diniz, 1968). Bradykinin releasing enzymes have been isolated from the venoms of *Agkistrodon halys blomhoffi* and *Echis coloratus.* Both are arginine esterases whose action is on an α-globulin precursor in normal plasma (Suzuki, 1966; Cohen et al., 1969). Bradykinin produces marked and sudden vasodilation accompanied by hypotension and smooth muscle stimulation that may be manifest as vomiting and diarrhea. It produces pain when injected locally. It does not have much lethal toxicity but may contribute to the rapid immobilization of prey.

Myonecrosis is characteristic of envenomation by many viperids but is also seen occasionally with cobra and other elapid venoms if not overshadowed by the more rapidly developing and lethal neurotoxic and cardiotoxic manifestations. It is most readily demonstrated in experimental animals if the lethal toxins are neutralized by antivenin which has little effect on the activity of the substances responsible for the myonecrosis caused by elapid venoms. Degeneration of all elements of the muscle fiber occurs (Stringer et al., 1971). The substances responsible for this injury have not been identified. Myonecrosis due to viperid venoms is practically always accompanied by local hemorrhage and damage to small arteries, so that some of the injury may be ischemic in nature (Homma and Tu, 1971). In rattlesnake venom myonecrosis, the sarcoplasmic reticulum and myofibrils were most vulnerable to injury; the mitochondria and sarcolemma were more resistant (Stringer et al., 1972). The very powerful myolytic activity of *Trimeresurus flavoviridis* venom depends on a heat labile protease acting in concert with a heat stable factor and phospholipase A (Kurashiga et al., 1966).

A number of phosphatases able to hydrolyze phosphate, ester, or anhydride bonds in nucleotides have been described from snake venoms. Most have pH optima of about 9 and occur as several isozymes. Phosphodiesterase I (Orthophosphoric diester phosphohydrolase) has been found in all snake venoms studied and is particularly high in those of *Bothrops jararaca, Crotalus adaman-*

teus and *Vipera russelii* among the viperids and *Hemachatus hae-machatus* among the elapids. It is one of three polynucleotidases, the others being Deoxyribonucleate 3'-nucleotidhydrolase (DNase) and Polyribonucleotide 2-oligonucleotidotransferase (cyclizing) (RNase). These also occur in all venoms, that of *Bitis gabonica* being a good source of both. Another very active and widely distributed phosphatase is 5'Ribonucleotide phosphohydrolase (AMPase) which is present in high concentrations in *Crotalus horridus, Bothrops atrox* and *Bitis gabonica* venoms. Adenosine triphosphatase and nucleotide pyrophosphatase activities have been reported for many snake venoms. ATPase occurs in high concentration in venoms of *Vipera russelii, Bothrops atrox,* and *Trimeresurus flavoviridis.* Its activity can be differentiated biochemically and biologically from that of phosphodiesterase (Pereira Lima et al., 1971).

A marked fall in blood pressure and other toxic effects have been reported following injection of snake venom phosphodiesterase and AMPase; however, the preparations used may have contained other toxins in close association with the enzymes (Russell, 1967). At present the role these enzymes play in envenomation and in the physiology of venomous snakes remains obscure.

Phosphomonesterases (Orthophosphoric monoester hydrolases) occur mostly in venoms of elapids such as *Naja naja* and *Naja haje* but are also found in fairly high concentration in the pit viper *Trimeresurus wagleri* whose venom is aberrant in some other respects. They have not been detected in sea snake venoms.

L-amino acid oxidase is an oxidoreductase capable of acting on a wide variety of substrates to catalyze the conversion of L-amino acids into α-keto acids. Its content of flavin adenine dinucleotide gives it a yellow color and permits its presence to be roughly estimated by visual inspection. In *Crotalus adamanteus* venom, it occurs as three isoenzymes; in *Agkistrodon piscivorous* it has a single molecular form (Wellner and Hayes, 1968). It is present in venoms of most species of viperids and elapids, although generally in lower concentration in the latter family; it has not been reported in hydrophiid venoms. It appears to be quite consistently lacking in venoms of very young snakes (Jimenez-Porras,

1964; Minton, 1967; Fiero et al., 1972). The activity begins to appear at 60-90 days of age. White or colorless venom has also been noted as a characteristic of certain subspecies of snakes, e.g. *Pseudocerastes persicus fieldi* or snakes of certain geographic areas such as *Vipera aspis* in the Department du Gers of France (Zeller, 1948). It may also be seen as an aberration of individual animals or the secretion of one gland may be white while the other secretes normally colored venom (Deoras, 1963). L-amino acid oxidase has been reported absent or in trace amounts in white venoms which frequently are more toxic than those containing large quantities of this enzyme.

L-amino acid oxidase has generally been assumed to play a part in the activity of snake venom as a digestive secretion; however, there is practically no direct confirmation of this. The enzyme does not contribute to the cardiovascular effects of snake venom, nor does it affect neuromuscular transmission. Its intrinsic toxicity for mammals is very low (Russell et al., 1963).

Hyaluronidase was one of the first enzymes to be definitely identified in venoms (Duran-Reynals, 1936) and its presence has been detected in many venomous animals, both invertebrates and vertebrates. It apparently occurs in nearly all snake venoms, although it was not detected in that of *Naja nigricollis* (Zwisler, 1966). This may be due to individual or geographic variations, for the enzyme has been reported in venom of this species by others (e.g. Schwick and Dickgiesser, 1963) albeit in low concentration. Very high activity was reported for the venom of *Bungarus caeruleus* (Jaques, 1956). Variable results have been obtained with other species either indicating considerable intraspecific variation or instability of the enzyme in venom preparations. It has not been possible to obtain highly purified hyaluronidase from snake venoms.

Hyaluronidase depolymerizes hyaluronic acid and dissolves the gel surrounding normal cells. It has virtually no intrinsic toxicity but facilitates the diffusion of other venom components through tissues.

Cholinesterase (Acetylcholine acetyl-hydrolase) is present in high concentration in most elapid venoms and was once believed to be responsible for their neurotoxic activity. It is lacking or present in only trace amounts in venoms of vipers and pit vipers including those such as *Crotalus durissus terrificus* venom which have a strongly neurotoxic activity. It is also absent in some elapid venoms, notably those of the mambas *(Dendroaspis)*. The snake venom cholinesterases are able to hydrolyze a variety of acetic esters. They may be a mixture of several isoenzymes (Boquet, 1964). The purified enzyme is heat labile losing 50 per cent of its activity at 44° in an hour. It is also inhibited by mild oxidizing agents, eserine, and ultraviolet light (Sarkar and Devi, 1968).

Certain snake venoms have long been known to have potent anticomplementary activity. Recently this activity was found in venoms of 18 of 32 species of reptiles. Venoms of seven species of cobras induced selective depletion of the C3-C9 sequence with the formation of a stable intermediate compound consuming these fractions. Venoms of seven other snakes (three species of *Agkistrodon,* three species of *Bothrops,* and *Bitis arietans*) acted on the C3-C9 sequence and also consumed C2 and C4 but without the formation of a stable intermediate. Other patterns of complement inactivation were shown by venoms of *Trimeresurus popeorum, T. purpureomaculatus, Lachesis mutus,* and the lizard, *Heloderma horridum* (Birdsey et al., 1971). A factor (CoF) which specifically inactivates C3 has been prepared from venom of the Asian cobra *(Naja naja)*. It has been characterized as a glycoprotein with the electrophoretic mobility of a β-globulin. Its action requires the presence of a thermolabile serum factor, evidently a 5S β pseudoglobulin, and bivalent cations (Müller-Eberhard and Fjellström, 1971). Injection of purified CoF depleted animals of 90 per cent of their complement activity, C3 being especially diminished. The complement level remained low for four to six days, but no serious side effects were observed. Immunologic reactions requiring participation of neutrophils, e.g. nephrotoxic nephritis and Arthus reactions, were inhibited; reactions not requiring neutrophil participation such as cutaneous anaphylaxis were unaffected. There was no effect on the induction of delayed hypersensitivity using tuberculin or synthetic antigens (Cochrane et al., 1970; Schwartz and Naff, 1971).

Polypeptide Toxins

Despite the plethora of enzymes in snake venoms, most of the lethal effects appear to be due to polypeptides and low molecular weight proteins without enzymatic activity. The first lethal factor to be isolated was crotoxin from the Brazilian rattlesnake, *Crotalus durissus terrificus* (Slotta and Fraenkel-Conrat, 1938). Crotoxin eventually proved to be a mixture of phospholipase A and two polypeptides which are the toxic moieties. One of these, crotactin, is of general occurrence in rattlesnakes of the *Crotalus durissus* complex; the other, crotamine, does not occur in the venom of all tropical American rattlesnakes but is limited to those of certain geographic regions (Goncalves, 1956; Schenberg, 1959). Both these toxins produced paralysis and respiratory distress in animals.

The first reasonably pure elapid toxins were obtained from venom of the Indian cobra (Ghosh et al., 1941) and the Cape cobra, *N. nivea* (Micheel et al., 1937); their effects were neurotoxic. A toxin with direct effect on cardiac muscle was isolated from Indian cobra venom (Sarkar, 1947). Crystalline neurotoxin (cobrotoxin) was prepared from venom of the Chinese cobra by Yang (1965) and subsequently at least a dozen additional cobra neurotoxins have been isolated (Table VII). Other neurotoxins have been isolated from venoms of sea snakes, kraits, other elapids, and a few viperids.

Of the elapid and sea snake neurotoxins that have been adequately characterized, all but two are composed of a single chain of 61 to 74 amino acids cross-linked by four or five disulfide bonds. There is a high degree of similarity in amino acid composition and sequence. Basic amino acids such as lysine and arginine seem to be essential for biological activity. Most of these neurotoxins accumulate at the motor endplate zone and produce an anti-depolarizing neuromuscular block without inhibition of acetylcholine output from motor nerve endings. The effect on neuromuscular transmission is reported to be much like that of d-tubocurarine (Lee, 1970).

A second type of elapid neurotoxin is exemplified by β-bungarotoxin from the krait, *Bungarus multicinctus*. This is a larger molecule of considerably different amino acid composition. It has a presynaptic action at the neuromuscular junction and inhibits the release of acetylcholine while leaving the sensitivity of the motor endplate to acetylcholine unimpaired. It has a longer latent period than cobra neurotoxins and α-bungarotoxin (Chang and Lee, 1963). Notectin recently obtained from tiger snake *(Notechis scutatus)* venom appears to be similar in mode of action. It occurs with at least three other neurotoxins, two of which are of the shorter chain type with postsynaptic site of activity (Karlsson et al., 1972).

Viperatoxin differs from elapid neurotoxins in amino acid composition and in showing a lower degree of cross-linkage. Its primary action is on the medullary centers producing vasodilatation and cardiac failure.

Crotoxin produces a non-depolarizing neuromuscular block accompanied by a decrease in the sensitivity of the motor endplate to acetylcholine (Vital Brazil, 1966). It also is nephrotoxic, but this probably is associated with the presence of phospholipase A in the toxin (Hadler and Vital Brazil, 1966). Crotamine produces muscular contracture that has been likened to that caused by veratrum alkaloids. This is evidently due to a direct effect of crotamine on skeletal muscle. It may be followed by a secondary paralysis (Cheymol et al., 1966).

Low molecular weight basic protein toxins have been isolated from venoms of several North American rattlesnakes. Both chemically and in their pharmacological activities they resemble crotamine. They contain 38-61 amino acid residues. In mice they produce immediate prostration with extreme tonic hyperextension of the hind limbs, exophthalmos, and gasping respirations but no trace of hemorrhage or necrosis. They make up 2 to 8 per cent of the venom protein of *Crotalus adamanteus, C. horridus,* and some, but not all, subspecies of *C. viridis.* In eight additional rattlesnake species and in four pit vipers of the genera *Agkistrodon* and *Bothrops,* these toxins were lacking or found only in concentrations of less than 1 per cent (Bonilla et al., 1971; Bonilla and Fiero, 1971). A toxin with a molecular weight of less than 1000 as determined by membrane filtration and high heat stability (125°

TABLE VII. SOME POLYPEPTID TOXINS OF SNAKE VENOMS

Name	Source	M.W.	Amino acid Residues	Mouse LD_{50} mg/kilo	Remarks	References
Toxin A	*Naja naja naja*		61	0.15		Nakai et al, 1971
Toxin 3	” ” ”	7834	71	0.10		Karlsson & Eaker 1972
Oxiana toxin	*Naja n. oxiana*	6786	61	0.10		” ”
Cobrotoxin	*Naja n. atra*	6949	62	0.05		Yang et al, 1969
Neurotoxin α	*Naja nivea*	7897	71			Botes et al, 1971
” β	” ”	6973	61			” ”
” δ	” ”	6834	61			” ”
Toxin α	*Naja nigricollis*	6787	61	0.09		Karlsson et al, 1966
Toxin α	*Naja haje*	6834	61	0.105	Identical with *Naja nivea* Neurotoxin	Botes & Strydom, 1969
Hemachatus II	*Hemachatus haemachatus*	6838	61	0.11		Strydom & Botes, 1971
Hemachatus IV	”	6831	61	0.09		” ”
α Bungarotoxin	*Bungarus multicinctus*	7983	74	0.30		Chang & Lee, 1963; Lee, 1970
β Bungarotoxin	”	28,500	179	0.089		” ”
Toxin b	*Naja melanoleuca*		71		Identical with *Naja nivea* Neurotoxin	Botes, 1972
Toxin d	” ”		61		Similar to *Hemachatus* toxins II and IV	
Notectin	*Notechis scutatus*	13,574	119	0.025		Karlsson et al, 1972
Mamba α	*Dendroaspis polylepis*	6907	60	0.09	4 tyrosyl residues	Strydom, 1972
Mamba γ	” ”	?	72	0.12	2 tryptophanyl residues	”

Erabutoxin a	*Laticauda semifasciata*	6837	62	0.15	Very similar toxins, isolated from *L. laticaudata* (laticotoxin) and *L. colubrina*	Tamiya & Arai, 1966; Lee, 1970
Erabutoxin b	"	6857	62	0.15		" "
Enhydrina Toxin	*Enhydrina schistosa*	6981	62	0.025		Tu & Toom, 1971 a
Lapemis Toxin	*Lapemis hardwickii*	6774	61			" 1971 b
Viperatoxin	*Vipera x. palaestinae*	12,212	108	0.30		Moroz et al, 1966
Crotamine	*Crotalus d. terrificus*	5450	46	0.07		Goncalves & Giglio, 1964
Crotactin	"	8582		0.06		Habermann & Rubsamen, 1971
Adamanteus Neurotoxin	*Crotalus adamanteus*	10,900	83		Ca $\frac{1}{4}$ as toxic as crude venom	Bonilla et al, 1971

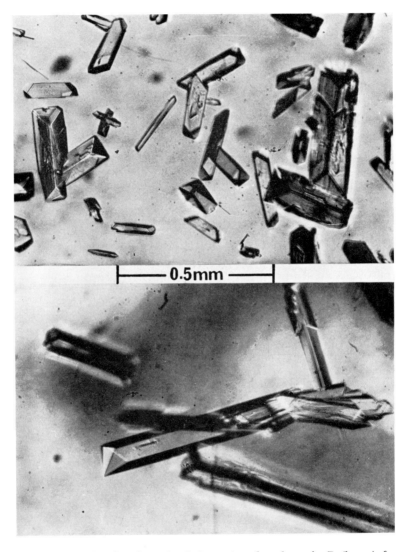

Figure 36. Crystals of erabutoxin A (upper) and erabutoxin B (lower) from venom of the sea snake *Laticauda semifasciata*. Both are neurotoxins with 62 amino acid residues and molecular weights of about 6800. (Preparation and photo courtesy N. Tamiya.)

for 15 min) has been obtained from the venom of the pit viper, *Trimeresurus wagleri*. In mice it produces rapid collapse and hyperpnea without hemorrhage or necrosis (Minton, 1968a).

Although a cardiotoxic component of cobra venom was first separated more than two decades ago, it has only recently been obtained in highly pure form. It is the most basic of the cobra venom toxins and highly resistant to heat and tryptic digestion. Its molecular weight is comparable to that of the elapid neuro-toxins, but it is quite different in chemical composition. It has been shown to be identical with the direct lytic factor (Slotta and Vick, 1969) and probably with cobra cytotoxin and cobramine B (Jimenez-Porras, 1970). Cardiotoxin appears to be a broad spectrum cell membrane poison with diverse pharmacological effects, but its basic action is to produce irreversible depolarization of cell membranes. Its action is potentiated by phospholipase A (Lee et al., 1968). Its structure and activity have been likened to that of mellitin, but this may be oversimplification. The best character-ized cardiotoxins have come from Asian cobras; however, direct lytic factors occur in venoms of other cobras and cobra-like elapids. That from *Hemachatus* has been purified and characterized (Aloof-Hirsch et al., 1967).

Low Molecular Weight Compounds

The presence of large amounts of acetylcholine in the venom of mambas *(Dendroaspis)* is noteworthy and probably correlated with the absence of acetylcholinesterase in the venoms of these snakes. Venoms of snakes of ten other genera either contained only small amounts of acetylcholine-like substances or were completely negative. It has been suggested that the action of acetylcholine may be one of producing pain and thus primarily defensive. It may also act directly on the heart or neuromuscular junction (Welsh, 1967). Mambas are among the fastest of snakes and large-ly arboreal. They are reported to be aggressive hunters that take active prey such as birds, and the advantage of a rapidly acting toxin such as acetylcholine would be considerable.

Although serotonin and related tryptamine compounds are important constituents of many venoms, they are relatively unimportant in snake venoms. In twenty snake venoms tested, Welsh (1966) found serotonin in concentrations of greater than 5 mcg/gm only in the venoms of *Bitis gabonica,* a very large African viper, and *Sistrurus miliarius,* a small rattlesnake.

ANTIGENIC RELATIONSHIPS AND NATURAL IMMUNITY

With their high content of enzymes and other antigenic substances, snake venoms show complex patterns when studied by immunodiffusion or immunoelectrophoresis. Immunologic analysis of venoms has been used as a technique to investigate relationships within genera, among families, and among populations and individuals of the same species. In general, antigenic differences in venom tend to reflect differences in morphology at the generic and family levels. There is comparatively little sharing of antigens between elapid and viperid venoms; however, *Echis carinatus* has at least two antigens in common with *Naja nigricollis,* and one of these is shared with other species of *Naja, Hemachatus* and *Walterinnesia. Naja haje* and *N. nigricollis* antivenins neutralized moderate amounts of *Echis* and *Cerastes* venoms (Boquet et al., 1966; Mohamed et al., 1973a, b). There is considerable sharing of antigens between pit vipers and other viperids (Puranananda et al., 1966; Minton, 1967 and unpublished). The venom of *Causus rhombeatus,* a primitive viper, gives precipitin bands with a number of elapid and viperid antisera as well as with *Dispholidus,* a colubrid antiserum. The weak immunological reactions between mole viper *(Atractaspis microlepidota)* venom and viperid antisera support the separation of atractaspids from the Viperidae (Minton, 1968b). Sea snake venoms are immunologically related to those of elapids, but only a limited amount of work has been done with them. Very weak cross reactions occur between *Dispholidus* venom and venoms of a number of pit vipers as well as that of one elapid *(Pseudechis papuanus).* The precipitin bands are seen only after preparations are stained with amidoschwarz or similar stains (Minton, unpublished). There is no information on other colubrid venoms.

At the species and subspecies level, antigenic differences are generally less marked and more difficult to interpret. Species such as *Agkistrodon rhodostoma* and *Trimeresurus wagleri* that are morphologically distinctive in their respective genera also have immunologically distinctive venoms. However, antigenic differences between venoms of newborn rattlesnakes and that of their mother were greater than those observed between adults of several species of rattlesnakes (Minton, 1967). Schenberg (1963) found 20 immunological variants in study of *Bothrops neuwiedi* samples from 298 snakes. Most of these were slight and could not be correlated with morphological differences. On the other hand, I obtained markedly different immunoelectrophoretic patterns between venoms of two closely-related horned vipers, *Pseudocerastes p. persicus* and *P. p. fieldi*.

Immunodiffusion techniques have shown common antigens in snake venoms and snake sera (Tu and Gantavorn, 1968; Minton, 1973). The significance of these is questionable. They evidently do not account for the toxicity of snake sera, for this toxicity is not neutralized by antivenin (Latifi and Shamloo, 1965).

Immunodiffusion techniques have not proved particularly useful in predicting the therapeutic effectiveness of antivenins except

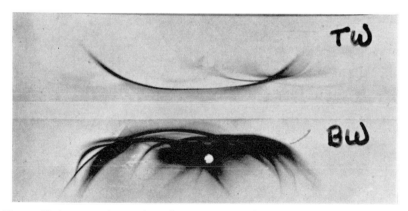

Figure 37. Immunoelectrophoresis preparation of venoms of Palestine Horned Viper *(Pseudocerastes persicus fieldi)* in upper well and Persian Horned Viper *(P. p. persicus)* in lower well. Preparation run 60 min. at 7.4 v/cm and developed with *P. p. persicus* antivenin. Although these two snakes are considered geographic races of the same species, there is a marked antigenic difference between their venoms. (Photo by Illustration Department, Indiana University Medical Center.)

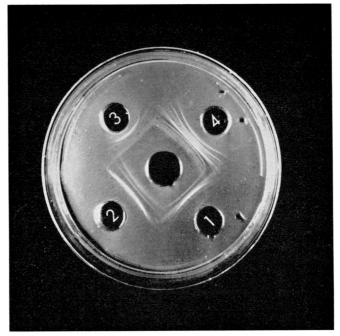

Figure 38. Antigenic differences among venoms of four pit vipers of the genus *Trimeresurus*. Well 1: *T. okinavensis;* Well 2: *T. albolabris;* Well 3: *T. wagleri;* Well 4: *T. flavoviridis;* Center well: *T. flavoviridis* antivenin. (Photo by Illustration Department, Indiana University Medical Center.)

where precipitin lines have been identified as due to specific toxins. Some theoretical aspects of the immunoelectrophoresis of snake venoms have been discussed by Jouannet (1968).

The relative immunity of venomous snakes to their own venom and that of related species has long been known. Rattle-snakes and other North American pit vipers possess much greater resistance to crotalid venoms than mammals, although naturally inflicted bites are fatal under some circumstances. *Crotalus adamanteus* serum, after heating at 56° to inactivate its toxicity, protected mice against the homologous venom. This activity was associated with the albumin fraction of the snake serum rather than the immunoglobulins (Clark and Voris, 1969). Serum of *Vipera berus* similarly was effective in neurtalizing homologous

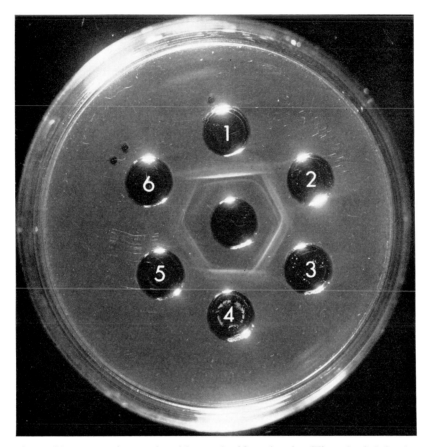

Figure 39. Preparation made with an elapid antivenin (Tiger Snake) in the center well and venoms of six species of vipers in the peripheral wells. Well 1: *Echis carinatus;* Well 2: *Crotalus atrox;* Well 3: *Agkistrodon contortrix;* Well 4: *Bothrops atrox;* Well 5: *Trimeresurus flavoviridis;* Well 6: *Vipera xanthina.* The presence of several precipitin lines indicates sharing of venom antigens between Tiger Snake and several species of vipers. (Photo by Illustration Department, Indiana University Medical Center.)

venom but not that of the related *V. aspis.* In this instance, however, neutralization was associated with the globulin fraction of the snake serum (Boquet, 1945). Tiger snakes *(Notechis scutatus)* proved very highly resistant to their own venom and that of some other elapids; the mechanism of the immunity was not investigated but it was believed higher than could be accounted for by factors in the serum (Kellaway, 1931).

Natural immunity to venom has been observed in snakes which pray upon poisonous snakes such as the kingsnakes *(Lampropeltis getulus)* of the U. S. Kingsnake serum proved moderate-

Figure 40. Presence of antibodies in serum of individual immunized against venom of the Asian Cobra *(Naja naja)*. His serum is in the center well. Wells 1, 3, and 5 contain *Naja naja* venoms from three different sources. Well 2 contains *Echis carinatus* venom, Well 4, *Vipera russelii* venom, and Well 6 *Bungarus caeruleus* venom. No detectable antibody to venoms other than that of *Naja naja* has been produced. Presence of antibodies to whole *Naja naja* venom does not necessarily indicate a high degree of immunity to specific toxins in the venom. (Photo by Illustration Department, Indiana University Medical Center.)

ly effective in protecting animals against pit viper venoms (Philpot and Smith, 1950). The protective factor was found in the γ-globulin fraction of the serum and behaved as a typical antibody. It acted, at least in part, by inhibiting venom proteases (Bonnett and Guttman, 1971). Similar venom neutralizing activity was found in the serum of *Elaphe quadrivirgata,* a nonvenomous snake which is not known to prey upon venomous species (Philpot, 1954). Attempts to immunize snakes, including kingsnakes and rattlesnakes, with rattlesnake venom did not result in the production of detectable antibodies (Juratsch and Russell, 1971).

Immunity in certain mammalian predators of poisonous snakes has been reported (Grasset et al., 1935; Vellard, 1949); however, its mechanism has not been adequately investigated.

Chapter 6

CLINICAL ASPECTS OF SNAKEBITE

EPIDEMIOLOGY

THERE ARE NO RECENT figures for world morbidity or mortality from snakebite. Swaroop and Grab (1954) estimated a world mortality of 30,000 to 40,000 annually and, assuming a mortality of about 15 per cent, an annual incidence of about 300,000 bites. The latter figure is almost certainly too low; however, it is equally probable that the world mortality figure is now considerably too high.

The two general regions of highest snakebite incidence and mortality are southeast Asia and tropical America. Earlier data from these regions has been summarized by Minton and Minton (1969, pp. 68-77). A more recent although not comprehensive survey of snakebite in southeast Asia is that of Sawai et al. (1971). They reported the following figures for periods of approximately five years between 1965 and 1971: Malaysia: 5387 cases, 18 deaths; Thailand: 14,578 cases, 191 deaths; Philippines: 304 cases, 26 deaths; Taiwan: 335 cases, 17 deaths. These statistics were derived almost entirely from hospitalized cases. There are no recent statistics for any extensive region of tropical America; however, the Hospital Vital Brazil of the Instituto Butantan during 1954-1965 treated 1718 cases with 30 deaths (Rosenfeld, 1971).

Information on snakebite in the United States has come mostly from reports by Parrish (1957a, 1966). These indicate 6000 to 7000 cases annually with 14 to 15 deaths. The incidence of bites is highest in North Carolina; however, most of the fatalities have been in Arizona, Texas, Georgia, and Florida.

Information on the epidemiology of snakebite in Africa is sparse and fragmentary. The best summary is that of Chapman

142

(1968) which deals largely with South Africa. In Natal during a seven-year period he reported 1067 bites of which 9.9 per cent were severe and 2 per cent fatal. Reported mortality from Africa generally is low considering the large number of dangerous species of snakes native to the continent.

Although mortality is very low, the incidence of snakebites in Europe is moderately high even as far north as Finland where 163 bites were reported in the summer of 1961 (Tallqvist and Oster-lund, 1962). It is generally assumed that the incidence of both snakebites and fatalities is higher in southeastern Europe; however, there is little information available. Between 1951 and 1953, Israel reported 303 snakebite cases with 24 deaths and between 1955 and 1957, 418 cases with 17 deaths (Leffkowitz, 1962). There is practically no information on recent incidence of snakebite in other countries of the Near and Middle East.

There has been no recent comprehensive survey of snakebite incidence in Australia and Melanesia. Mortality is believed to be low, although a high percentage of the snake species in this region of the world are venomous. Trinca (1963) reported 45 deaths from snakebite in Australia in the 1952-1961 period. Eighteen of these occurred in Queensland and 9 each in New South Wales and Victoria.

A disproportionately high percentage of snakebites occur among persons deliberately handling venomous snakes for a variety of motives—bravado, showmanship, avocational, or scientific interest. Snake handling as a part of religious ritual is practiced in several diverse parts of the world (Minton and Minton, 1969, pp. 131-148, 176-187). Snakes, including venomous species, are used for food and in folk medicine in some cultures, and persons catching them for such uses may be bitten. If these special, high-risk groups are excluded, most snakebites occur among agricultural workers, herdsmen, fishermen, and those clearing land by largely manual methods. Children make up a high percentage of snakebite casualties and incidence is also high among adolescent and young adult males. Snakebites are infrequent among the elderly, but the proportion of fatalities is high.

In temperate zone countries, the incidence of snakebite is

largely restricted to the warmer months of the year when the
snakes are not in hibernation; exceptions are nearly always due to
bites of captive reptiles. The warm months are also generally
times of increased human outdoor activity. In semiarid tropical or
subtropical regions and in deserts, the peak of snakebite incidence
usually coincides with the rains and corresponds with a general
increase in animal and human activity. In the great river valleys
of the tropics and subtropics, snakebite incidence is often highest
during seasonal inundations that concentrate both snakes and the
human population on islands of high ground. In tropical regions
without strongly marked wet and dry seasons, snakebites are likely
to be more or less evenly distributed throughout the year. With
regard to time of day, most snakebites seem to be related to peaks
of human rather than reptilian activity. In arid and tropical
regions, the great majority of snakes are nocturnal or crepuscular,
yet most bites occur during daylight hours often as a result of dis-
turbing a snake in its resting place.

Some venomous snakes such as the cobras and kraits of southeast Asia adapt well to living in close proximity to man and may
enter houses in search of rodents and lizards. Ethnic factors that
may influence the incidence of snakebite include agricultural and
fishing practices, going barefoot or wearing inadequate footgear,
and sleeping on mats on the ground.

Major venomous snakes of various parts of the world are listed
in Table VIII. Additional information may be found in the U. S.
Navy publication *Poisonous Snakes of the World* (1968).

TABLE VIII. POISONOUS SNAKES OF MAJOR MEDICAL IMPORTANCE

United States and Canada
    Diamondback rattlesnakes (*Crotalus adamanteus* and *C. atrox*)
    Timber rattlesnake (*C. horridus*)
    Prairie rattlesnake (*C. viridis viridis*)
    Pacific rattlesnake (*C. viridis helleri* and *C. v. oreganus*)
    Pigmy rattlesnake (*Sistrurus miliarius*)
    Copperhead (*Agkistrodon contortrix*)
    Cottonmouth (*A. piscivorus*)
Mexico, Central America, and West Indies
    Western diamondback rattlesnake (*Crotalus atrox*)
    Mexican west-coast rattlesnake (*C. basiliscus*)
    Tropical rattlesnake (*C. durissus*) several subspecies

    *Barba amarilla or terciopelo *(Bothrops atrox)*
    Eyelash viper *(B. schlegelii)*
   **Fer-de-lance *(B. lanceolatus* and *B. caribbaeus)*
Northern South America (to about 15° S.)
    Tropical rattlesnake *(Crotalus durissus)* several subspecies
    Barba amarilla or terciopelo *(Bothrops atrox)*
    Amazonian tree viper *(B. bilineatus)*
    Bushmaster *(Lachesis mutus)*
    Amazonian coral snake *(Micrurus spixii)*
Southern South America
    Brazilian rattlesnake *(Crotalus durissus terrificus)*
    Jararaca *(Bothrops jararaca)*
    Jararacussu *(B. jararacussu)*
    Jararaca pintada *(B. neuwiedi)*
    Urutu *(B. alternatus)*
    Brazilian giant coral snake *(Micrurus frontalis)*
Europe
    European viper *(Vipera berus)*
    Asp viper *(V. aspis)*
    Long-nosed viper *(V. ammodytes)*
Near and Middle East
    Palestine or Turkish viper *(Vipera xanthina)*
    Levantine viper *(V. lebetina)*
    Saw-scaled vipers *(Echis coloratus* and *E. carinatus)*
Indian Subcontinent and Ceylon
    Russell's viper *(Vipera russelii)*
    Saw-scaled viper *(Echis carinatus)*
    Indian green tree viper *(Trimeresurus gramineus)*
    Indian krait *(Bungarus caeruleus)*
    Indian cobra *(Naja n. naja* and *N. n. kaouthia)*
    Sea snakes, especially the beaked sea snake *(Enhydrina schistosa)*, are
        important in some coastal areas.
Southeast Asia including Philippines and most of Indonesia
    Russell's viper *(Vipera russelii)*
  ***Malay pit viper *(Agkistrodon rhodostoma)*
    White-lipped tree viper *(Trimeresurus albolabris)*
    Mangrove viper *(T. purpureromaculatus)*
    Wagler's pit viper *(T. wagleri)*
    Malayan krait *(Bungarus candidus)*
    Asian cobras (Various races of *Naja naja,* chiefly *N. n. kaouthia, N. n.*
        atra, N. n. sputatrix, and *N. n. philippiensis)*
    King cobra *(Ophiophagus hannah)*
    Beaked sea snake *(Enhydrina schistosa)*
    Annulated sea snake *(Hydrophis cyanocinctus)*
    Hardwicke's sea snake *(Lapemis hardwickii)*
Far East (Most of eastern China, Korea, Taiwan, Japan)
    Hundred-pace snake *(Agkistrodon acutus)*

Mamushi (*Agkistrodon halys blomhoffi* and *A. h. brevicaudus*)
Okinawa habu (*Trimeresurus flavoviridis*)
Chinese habu (*T. mucrosquamatus*)
Chinese green tree viper (*T. stejnegeri*)
Many-banded krait (*Bungarus multicinctus*)
Chinese cobra (*Naja naja atra*)
Annulated sea snake (*Hydrophis cyanocinctus*)

New Guinea, Northern Australia and associated islands
Death adder (*Acanthophis antarcticus*)
Brown snake (*Pseudonaja textilis*)
Taipan (*Oxyuranus scutellatus*)
Mulga snake (*Pseudechis australis*)
Papuan blacksnake (*P. papuanus*)

Southern Australia and Tasmania
Death adder (*Acanthophis antarcticus*)
Brown snake (*Pseudonaja textilis*)
Copperhead (*Denisonia superba*)
Australian blacksnake (*Pseudechis porphyriacus*)
Tiger snakes (*Notechis scutatus* and *N. ater*)

North Africa (to southern edge of Sahara)
Desert horned viper (*Cerastes cerastes*)
Saw-scaled viper (*Echis carinatus*)
Puff adder (*Bitis arietans*)
Sahara rock viper (*Vipera mauritanica*)
Northern mole viper (*Atractaspis microlepidota*)
Egyptian cobra (*Naja haje*)

Central and southern Africa
Saw-scaled viper (*Echis carinatus*)
Puff adder (*Bitis arietans*)
Gaboon viper (*B. gabonica*)
Night adder (*Causus rhombeatus*)
Cape cobra (*Naja nivea*)
Egyptian cobra (*N. haje*)
Spitting cobra (*N. nigricollis*)
Ringhals (*Hemachatus haemachatus*)
Green mambas (*Dendroaspis viridis* and *D. angusticeps*)
Black mamba (*D. polylepis*)
Boomslang (*Dispholidus typus*)

---

*Central American populations long known as *B. atrox asper* have been given full species status by some recent workers.

**Islands of Martinique and St. Lucia only. The name "fer-de-lance" 'is often applied to mainland populations of *Bothrops atrox*.

***The generic name is often spelled *Ancistrodon*. Some recent workers place this snake in a monotypic genus, *Calloselasma*.

SYMPTOMATOLOGY

The clinical course of snakebite is characterized by a high degree of variability and unpredictability. More than in other envenomations, there are multiple factors that involve both the biting snake and the bitten human. Most of these have been summed up by Klauber (1956, pp. 814-825). The two most important and largely unknown factors are the intrinsic toxicity of the venom for man and the amount injected. The former can be only roughly estimated by extrapolation from data obtained by animal experiment. Man appears to be more susceptible on a weight-for-weight basis than the mouse or dog but less susceptible than the pigeon. Data from primate experiments may be directly applicable to man, but there is little such information available. The quantity of venom injected may vary from none to, in theory,

Figure 41. Tropical Rattlesnake *(Crotalus durissus ssp.)*, a species with extensive range in South and Middle America. Its venom shows marked geographic variation and in many localities is predominantly neurotoxic. (Photo by John Tashjian.)

Figure 42. Prairie Rattlesnake *(Crotalus v. viridis)*, a common and moderately dangerous rattlesnake of the Great Plains region of the United States. (Photo by author.)

the full amount in the snake's glands. Experiments with vipers indicate that this is rarely, if ever, the case; the usual amount injected appears to be 10 to 15 per cent of the content of the glands (Kochva, 1960; Gennaro et al., 1961; Kondo et al., 1972). Elapids and colubrids whose biting technique is different from that of vipers may inject a higher percentage of their venom, but observations in the field and laboratory indicate they also tend to inject considerably less than the full content of their glands.

The observation that the snake appears to have control over the amount of venom injected and may frequently inject very little when biting defensively has obvious clinical implications. Evaluation of large numbers of snakebite cases from various parts

Figure 43. Malay Pit Viper *(Agkistrodon rhodostoma)* curled about its eggs. This species is an important cause of snakebites in rubber-growing areas of southeast Asia. An effective anticoagulant (arvin) has been prepared from its venom. (Photo by Andrew Koukoulis.)

of the world indicates that almost half result in little or no envenomation even though the bites may be inflicted by large snakes with highly toxic venom (Hadar and Gitter, 1959; Reid, 1963; Parrish et al., 1966; Campbell, 1967).

The variables affecting both the quantity and quality of venom injected plus the manifold actions of snake venoms that affect nearly every organ system make it difficult to summarize the symptomatology that may be encountered in man. Manifestations of poisoning readily elicited in experimental animals may seldom be seen in human envenomation, and the reverse is also true. Russell and Puffer (1970) comment: "More errors in clinical judgment have been made on the basis of oversimplified classifications of the

Figure 44. *Bothrops atrox,* known by a number of Spanish and Portuguese names and often called "fer-de-lance" in English publications, is typical of the large and dangerous terrestrial pit vipers of tropical America. They are often plentiful in cultivated land. The specimen illustrated was collected in eastern Mexico and represents the subspecies *asper.* (Photo by author.)

physiopharmacological properties of snake venoms than are generally appreciated . . . the physician faced with a case of snake venom poisoning is confronted with two or perhaps three or four serious disease states."

Nutritional status and intercurrent disease can significantly influence response to envenomation. The individual with a heavy hookworm infection may show much more severe intestinal bleeding when bitten by some species of snakes than the individual not so parasitized. Other conditions prevalent in the tropics such as malaria, schistosomiasis, scurvy, and liver diseases also contribute to a poor prognosis in those bitten by venomous snakes, especially vipers (Corkill et al., 1959). There is a strong psychic element that varies among individuals and cultures. Few people are emotionally neutral toward snakes, and even herpetologists may lose

Figure 45. Saw-scaled Viper *(Echis carinatus)*, a small snake of dry country in northern Africa and western Asia, in its characteristic defensive pose. Rough scales produce a sizzling noise when air-filled loops of the body are rubbed together. Venom of this snake is highly complex antigenically and seems to have exceptionally high toxicity for man. (Photo by author.)

much of their objectivity when bitten.

Notwithstanding the great variation in response that may be seen, certain manifestations of snake venom poisoning are characteristic enough that they have been recognized widely in folklore as well as in early medical writings.

In most viperid bites, including those by rattlesnakes and other pit vipers of North America, there is free bleeding from the fang punctures and immediate burning local pain. However, in some severe rattlesnake bites local numbness may precede the onset of pain, and in a few cases pain is never a prominent symptom.

Figure 46. Russell's Viper *(Vipera russelii)* is a large species common in agricultural land in south Asia. It contributes substantially to snakebite fatalities in the Indian subcontinent, Ceylon and Burma. (Photo by author.)

Figure 47. Gaboon Viper *(Bitis gabonica)*, a tropical African reptile, is one of the largest vipers reaching a length of 204 cm and weight of 9 kg. It is comparatively inoffensive and causes few snakebite accidents. (Photo by author.)

Figure 48. Tiger Snake *(Notechis scutatus)*, an abundant Australian elapid whose venom is one of the most toxic known. This snake is quick in its movements and can be quite dangerous. (Photo by author.)

Figure 49. The Asian Cobra *(Naja naja)* probably causes more fatalities than any species of snake because it has highly toxic venom and a very wide range in densely populated regions. The Oxus Cobra *(N. n. oxiana)* shown here is the most northerly of its numerous geographic races and occurs from northern Pakistan to southern parts of Russian Asia. Several other species of cobras inhabit Africa. (Photo by Allan Roberts.)

Figure 50. Indian Krait *(Bungarus caeruleus)*. Kraits are vividly marked, nocturnal snakes widely distributed in southern Asia. Most species are 1 to 1.5 m when adult. They are not aggressive snakes, but their venoms are extremely toxic. (Photo by author.)

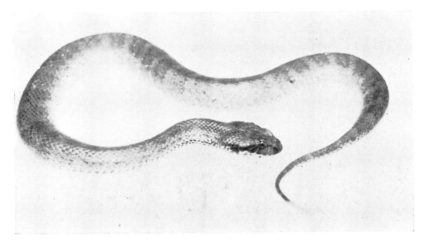

Figure 51. Death Adder *(Acanthophis antarcticus)*, an elapid that superficially resembles a viper. It is widely distributed in the Australia-New Guinea region. Although nocturnal and sluggish, it is a dangerous snake. (Photo by author.)

Figure 52. Hardwicke's Sea Snake *(Lapemis hardwickii)* is very plentiful in southeast Asian and northern Australian waters and is one of the more dangerous sea snakes. The broad, flat tail is characteristic of all sea snakes. (Photo by author.)

Figure 53. Annulated Sea Snake *(Hydrophis cyanocinctus)* is one of the larg-est sea snakes sometimes reaching a length of 2 m. It is found in Asian coas-tal waters from the Persian Gulf to the Sea of Japan. (Photo by author.)

Swelling usually begins promptly around the site of the bite and spreads centrally. This is often accompanied by discoloration of the skin and ecchymosis. Blood or serum-filled vesicles may appear within a few hours. The rapidity of appearance of these signs and their progression has been widely used in the United States for evaluating the severity of snakebites (Wood et al., 1955; McCol-lough and Gennaro, 1970); nevertheless, severe envenomation may occur with comparatively mild local manifestations. This is particularly true with some of the Old World and tropical Ameri-can viperids.

Early systemic symptoms include weakness, faintness, sweating, thirst, nausea and vomiting, and less frequently diarrhea. Tin-gling and numbness around the mouth and over the scalp are

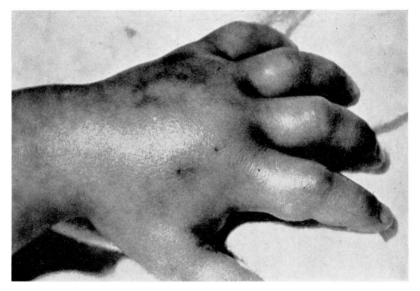

Figure 54. Rattlesnake bite of hand photographed soon after accident oc-curred; note punctures, swelling and discoloration. (Photo courtesy F. E. Russell.)

common with bites by rattlesnakes of the *Crotalus viridis* com-plex and occasionally seen with bites by other pit vipers. When bitten by a large viper, the patient may become unconscious with-in a few minutes after the bite but almost invariably regains con-sciousness within a short time. This may be in some cases an emo-tional reaction, but hypotension from bradykinin release is an-other possible explanation. Pain along the lymphatics and swell-ing of the regional lymph nodes is characteristic of North Ameri-can pit viper bites but infrequent with viper bites in southeast Asia. Moderate to profound hypotension may be seen, and the temperature is often elevated to 38-39°. Muscle fasiculations sometimes progressing to violent spasms or convulsions are seen with North American pit viper bites.

Hemorrhagic manifestations dominate the picture in bites by the lancehead vipers *(Bothrops)* of tropical America, many African and Asian vipers *(Echis* sp., *Vipera russelii, Agkistrodon rhodo-stoma,* and others) , the Australian blacksnake *(Pseudechis por-*

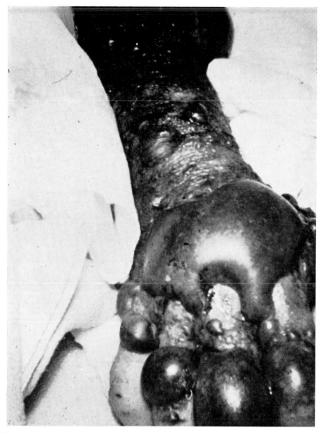

Figure 55. Severe bite by Timber Rattlesnake showing large hemorrhagic blisters on fingers and dorsum of hand. Serious systemic involvement. Photographed 4 days after injury. Recovery with minimal disability. (Photo courtesy F. E. Russell.)

phyriacus) and brown snake *(Pseudonaja textilis)* which are elapids rather than vipers, and the boomslang *(Dispholidus typus),* and bird snake *(Thelotornis kirtlandi),* African colubrids.

There is persistent oozing of blood from fang punctures. Hemoptysis following coughing may occur as early as 20 minutes after the bite and is a valuable diagnostic sign of systemic envenomation (Reid, 1968a). Bleeding from the gums is another early symptom. Ecchymoses may appear spontaneously on any part of

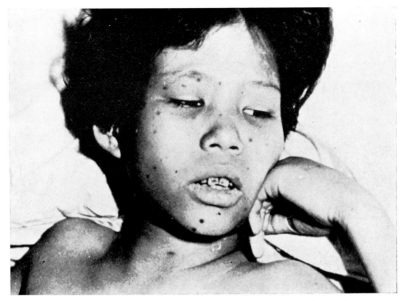

Figure 56. Discrete petechial hemorrhages in a patient with moderately severe envenomation from the bite of a Malay Pit Viper. (Photo courtesy H. A. Reid.)

the body or may be seen at sites of mild trauma or infection. Gross hematuria, hematemesis, vaginal bleeding, and blood in the feces may be observed. Cerebral hemorrhage may be fatal or be followed by temporary or permanent neurological deficit (Reid et al., 1963a; Ameratunga, 1972).

A prolonged clotting defect has been observed following bites by most of the above snakes as well as those by certain species of rattlesnakes *(Crotalus adamanteus, C. atrox, C. ruber, C. viridis)*. In Malay pit viper *(Agkistrodon rhodostoma)* envenomation, this coagulation abnormality may last as long as 26 days but frequently is relatively benign and accompanied by little evidence of internal or external hemorrhage (Reid et al., 1963b; Reid and Chan, 1968). Decreased fibrinogen and abnormal prothrombin times were seen in 10 of 13 children bitten by green tree vipers *(Trimeresurus)* in Thailand. Only 6 had clinically manifest hemorrhagic symptoms (Mitrakul and Impun, 1973). On the other hand, I have known

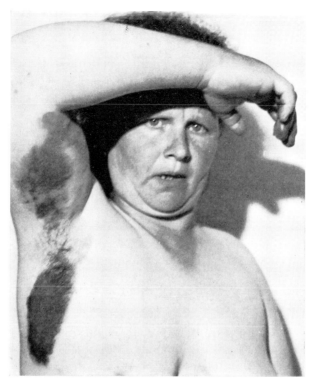

Figure 57. Ecchymoses of upper arm and axillary region following bite on finger by *Vipera ammodytes,* one of the more dangerous European vipers. (Photo courtesy Z. Maretic.)

patients bitten by the saw-scaled viper *(Echis carinatus)* to die on the 11th and 12th day after injury with evidence of intracranial hemorrhage. Evidence of pulmonary hemorrhage was noted in a fatal case of rattlesnake bite on the sixth day after injury; another patient with rattlesnake bite had bloody diarrhea for six days following injury (Minton, unpublished; Lyons, 1971). Widespread ecchymoses and other hemorrhagic phenomena were present in patients with boomslang envenomation (Mackay et al., 1969; Lakier and Fritz, 1969).

In instances of massive viper envenomation, death from shock may occur within 30 minutes. Most rattlesnake bite fatalities in the U. S. are seen between 6 and 30 hours after injury, and the same seems to be true of puff adder bites in Africa (Chapman, 1968) and Palestine viper bites in the Middle East (Efrati and Reif, 1953). In most of these fatalities, the chief finding at autopsy is massive extravasation of blood extending peripherally from the bitten area. Survival times following bites by the large *Bothrops* pit vipers of tropical America and the more dangerous vipers of Asia are more unpredictable. Death often results from massive or multiple hemorrhages incident to severe coagulopathy. Autopsy findings have been reported by Jutzy et al. (1953), Lieske (1963), and Chowhan (1938).

Clinically manifest intravascular hemolysis is uncommon following bites of most venomous snakes but is seen regularly in bites by the Brazilian rattlesnake *(Crotalus durissus terrificus)*. Methemoglobinuria and a nephrotic syndrome with lesions of the renal tubules is a particularly dangerous sequel (Amorim and Mello, 1954). Hemolysis has also been reported following bites by some of the Australian elapid snakes (Kellaway, 1938; Trinca 1963) and the saw-scaled viper (Weiss et al, 1973).

Necrosis around the bite and occasionally involving much of the injured limb is a fairly common sequel of viper bites. It is particularly severe following bites of the diamondback rattlesnakes *(Crotalus atrox* and *C. adamanteus)* in the U. S., some of the large lancehead vipers such as *Bothrops alternatus* and *B. jararacussu* in South America, the puff adder *(Bitis arietans)* in Africa, and the habus *(Trimeresurus flavoviridis* and *T. mucrosquamatus)* in the Far East. The necrosis around the fang punctures is due to the proteolytic action of the venom. More extensive necrosis is caused by interference with the blood supply to the part due to the pressure of subfascial edema or thrombosis of major vessels. Damage may be extensive requiring amputation or resulting in permanent impairment of function.

Bites of elapid snakes with certain exceptions, particularly among the Australian species, present a different clinical picture. Immediate pain around the bite is not so prominent a feature, although it may be fairly severe in cobra bites. Cobra bites also show a moderate degree of swelling. This is not so evident in bites by other elapids and may be completely lacking.

Persons bitten by large cobras or mambas may develop signs of severe systemic poisoning within 15-30 minutes. With kraits, coral snakes, and many other elapids, there is a latent period of 1 to 4 hours during which there are no symptoms of envenomation; in exceptional instances this may be considerably longer. One patient bitten by a coral snake showed no evidence of systemic envenomation for 22 hours but then developed severe paralysis requiring intensive therapeutic effort (Moseley, 1966). Early signs are drowsiness, weakness, difficulty in swallowing and speaking, and double vision. There may be some degree of mental confusion and euphoria. Severe abdominal pain may occur in krait and coral snake bites and may be accompanied by vomiting. Headache is a common early symptom in bites by some Australian elapids. Bilateral ptosis is usually the first objective sign of systemic elapid poisoning. The pupils are dilated but react normally to light. Increased salivation and sweating are frequently observed. Flaccid paralysis affects all muscle groups, although in no characteristic sequence, and is not accompanied by pain on passive movement or pressure. Breathing is shallow and diaphragmatic, and the blood pressure low. The temperature is usually subnormal. Coma, loss of deep tendon reflexes, and incontinence occur shortly before death. Despite the presence of a cardiotoxin in cobra venoms, cardiovascular changes are not a prominent part of the picture of elapid envenomation. Electrocardiographic abnormalities have been described in cobra envenomation, but may be in part secondary to respiratory failure (Reid, 1968b). In patients that recover from elapid envenomation, the neurological abnormalities resolve rapidly and completely.

Extensive necrosis following cobra bites has been reported in Malaya (Reid, 1964), elsewhere in southeast Asia, and in Africa (Davidson, 1970). It resembles that seen in viperid bites but is slower to manifest itself, often not being evident until two to five days after the bite. It may occur without signs of systemic poisoning or may be concomitant with it. The pathogenesis is not understood.

Sea snake bites cause little pain and no swelling at the site of injury. If significant envenomation has occurred, muscle stiffness

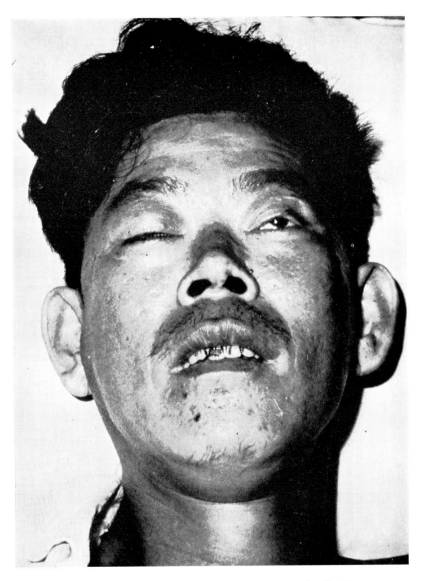

Figure 58. Ptosis is a characteristic early sign of systemic elapid envenomation. (Photo courtesy H. A. Reid.)

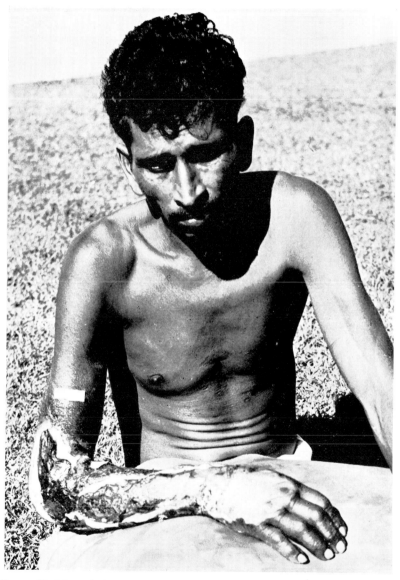

Figure 59. Necrosis is not uncommon following cobra bites and may or may not be accompanied by signs of systemic envenomation. Necrosis of the extent shown here requires several days to develop especially if corticosteroids have been given. The pathogenesis is poorly understood. (Photo courtesy H. A. Reid.)

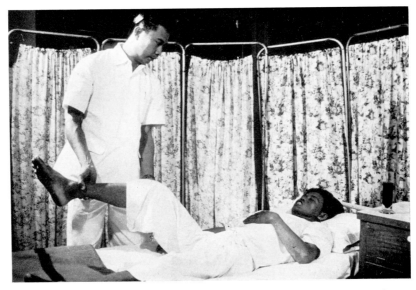

Figure 60. In patients with systemic envenomation by sea snakes, pain on passive movement of the limbs is usually seen within 30 to 60 minutes after the bite. (Photo courtesy H. A. Reid.)

and pain on passive movement are noted after a latent period of 15 minutes to 8 hours. This is accompanied by generalized muscular weakness. Trismus and ptosis are early signs of poisoning. Tendon reflexes, normal in the early stages of the illness, become progressively diminished. Myoglobinuria may be detected spectroscopically within a few hours after the bite and may be present when other signs of envenomation are minimal. In severe cases, the urine becomes reddish brown. Hyperkalemia with resultant cardiac abnormalities characterizes severe cases. Death from respiratory or cardiac failure usually occurs within 48 hours but may be delayed as long as 12 days. Seriously envenomed patients who recover may not regain full muscular strength for a several months and may show residual renal damage (Reid, 1957, 1961; Karunaratne and Panabokke, 1972). The most important primary lesion in sea snake poisoning is skeletal muscle damage, although a neurotoxic component in the venom may also contribute to the symptomatology. Most recent reports of sea snake bites have come from Malayan waters and most have been ascribed to the beaked seasnake *(Enhydrina schistosa)* which is the commonest species in that region. It is possible that bites by other species may show a somewhat different clinical picture.

Fatal snakebites accompanied by myoglobinuria, muscle pain, and trismus have been reported from Australia and ascribed to terrestrial species such as the small-eyed snake *(Cryptophis nigrescens)* and the mulga snake *(Pseudechis australis)*. These patients showed more local pain, headache, and vomiting than is seen with sea snake poisoning (Furtado and Lester, 1968; Rowlands et al., 1969).

Specific sensory disturbances as a direct result of snakebite are infrequent; however, a sensation of yellow vision has been reported as an early symptom of diamondback rattlesnake bite (Andrews and Pollard, 1953). Venom of the berg adder *(Bitis atropos)*, a small South African snake specifically affects some of the cranial nerves producing a rapid loss of taste, smell, and balance. There is marked ptosis and impairment of extraocular movements; the pupils are dilated and do not react to light or accommodation; and the corneal and gag reflexes are diminished or lost. Generalized muscular weakness is not marked. The venom produces some local pain and swelling, but clotting perimeters are unaffected. Cranial nerve functions may require 5 to 14 days to return to normal (Hurwitz and Hull, 1971; Visser, 1966).

Complete or partial blindness may follow bites of snakes such as the saw scaled vipers *(Echis)* and is attributed to optic atrophy secondary to hemorrhage or prolonged spasm of the retinal vessels (Davenport and Budden, 1953; Guttmann-Friedman, 1956).

Snakebites during pregnancy have been reported by Malz (1967), Parrish and Khan (1966) and Reid and Chan (1968). All bites were by vipers. Most of the women recovered and gave birth to healthy infants. In the group of 19 patients, three aborted spontaneously and one had a therapeutic abortion. One woman bitten in the first trimester of pregnancy gave birth to a child with multiple malformations. The likelihood of abortion seems related to the severity of envenomation rather than to the duration of the pregnancy.

A singular type of envenomation is the ophthalmia produced by the spitting cobras. In these snakes the discharge orifice of the fang is on the anterior aspect rather than at the tip, and powerful contraction of muscles surrounding the venom gland squirts jets of venom forward and slightly upward to distances of 2 m in the case of large snakes. The snakes seem to aim for strongly refractile objects which, under natural conditions, directs the venom toward the eyes. The most efficient "spitter" is the ringhals *(Hemachatus haemachates)* of southern Africa, but *Naja nigricollis* which ranges over a great part of Africa also utilizes this defense effectively. Cobra populations in Malaya, Indonesia, and the Philippines have fangs modified for "spitting" but the behavior pattern is not well developed. Ejection of venom is used only for defense, never toward the snakes' prey. These snakes may also bite as a defensive reaction.

Venom droplets striking the eye cause immediate intense pain and blepharospasm. There is marked conjunctival and corneal inflammation that usually subsides rapidly even without specific treatment. In one case there was a severe keratitis with corneal anaesthesia lasting about a week. A corneal opacity could be detected two months later, but vision was virtually normal (Ridley, 1944; Gilkes, 1959). Anaesthetized animals may be killed by instilling cobra venom into the eye, but systemic envenomation by this route has not been reported in man.

Snake venoms are allergenic, and sensitization may occur by injection, inhalation, and possibly other routes. Using scratch test and passive antibody transfer (modified Prausnitz-Küstner method), Parrish (1957b) demonstrated hypersensitivity to cottonmouth and diamondback rattlesnake venoms in seven of 30 persons with a history of one or more bites by North American pit vipers. Clinical allergy manifest as conjunctivitis, rhinitis, or asthma was found in 10 workers in laboratories where snake venoms (mostly *Bothrops* and *Crotalus*) were handled. Although most of these individuals had worked only with *Bothrops jararaca* venom, they demonstrated by skin tests and local passive transfer technique some degree of hypersensitivity to other viper venoms and, in one case, to coral snake venom. Several were sensitive to a therapeutic coagulant preparation derived from *Bothrops* venom. One individual subsequently had anaphylactoid symptoms following a jararaca bite (Mendes et al., 1960).

Lounsberry (1934) reported the case of a rancher who developed generalized urticaria and collapse within 5 minutes after the second of two rattlesnake bites suffered about a year apart. Symptoms of severe envenomation were followed by an exfoliative dermatitis lasting several weeks. A herpetologist with known sensitivity to rattlesnake venom suffered anaphylactic shock following the bite of a small rattlesnake (Ellis and Smith, 1965). While most reported cases of hypersensitivity have been to viperid venoms, a technician in my laboratory developed respiratory allergy to lyophilized cobra venom to such a degree she was unable to do further work with the material. She experienced no discomfort when working with viper venoms.

TREATMENT

Therapies for snakebite are as old as medicine itself, and their variety is limited only by the scope of human imagination. However, only a few are of proven value. Oldest of these are procedures designed to remove venom mechanically, exemplified today by various ligature, incision and suction techniques, or surgical excision of the area around the bite. Generally accepted first aid for bites by North American pit vipers is application of a constriction band proximal to the bite tightly enough to occlude superficial venous and lymphatic return but not arterial flow. It should be released for 90 seconds every 10 minutes and removed as soon as antivenin or other specific treatment is begun. In no case should it be left more than four hours. Two or three longitudinal incisions 5-7 mm long are then made through the full thickness of the skin but not into deeper tissues. These are best made through the fang punctures if these are identifiable. If, as is sometimes the case, there are additional punctures made by the snake's palatine or mandibular teeth, the incisions are made in the general area of the bite. Suction, preferably using a mechanical device, is applied for approximately an hour.

Experimental studies with ^{131}I labelled rattlesnake venom have shown that up to 50 per cent of the subcutaneously injected dose can be removed if suction is started within three minutes after injection. Even after two hours, 20 to 30 per cent of the labelled venom can be removed (Snyder et al., 1967; Gennaro, 1963). Bioassay of material obtained by suction from incisions into the tissues around rattlesnake bites has also shown the presence of appreciable amounts of venom (Merriam and Leopold, 1960; Russell and Emery, 1961). Both experimental and clinical evidence indicate incision and suction is of little value when begun more than 30 minutes after the bite. It has been observed that incision and suction is of little value in treatment of bites by elapid snakes whose low molecular weight toxins are very rapidly absorbed. The immediate application of a tight tourniquet after bites by large cobras, mambas, taipans, and other highly dangerous elapids has been recommended and has been shown to be effective in animals (Christensen, 1969). It is not removed until antivenin has been administered. This is obviously a drastic measure that risks permanent damage to a limb in order to save life. Incisions are contraindicated in patients bitten by saw-scaled vipers and other snakes whose venoms produce prolonged and severe coagulopathy.

Excision of tissue around the fang wounds after application of a ligature has been shown to remove as much as 90 per cent of ^{131}I labelled rattlesnake venom if carried out within 35-40 minutes. The procedure was used with good results in 32 patients bitten by rattlesnakes or cottonmouths (Snyder, 1966). Glass (1973) recommends prompt surgical exploration of pit viper bites with debridement of devitalized tissue. In his series of 84 cases, 58 per cent had evidence the venom was injected intramuscularly.

Despite certain shortcomings and very real hazards, antivenin (serum of animals immunized against venom or its fractions) is the most effective therapy for life endangering snakebite. Antivenins are currently produced by at least 27 laboratories in 24 nations (Russell and Lauritzen, 1966). There are no international standards, and products vary considerably in antibody content, degree of refinement, pyrogenicity, shelf life, and other properties. Antivenins may be monovalent or polyvalent against venoms of snakes of a particular geographic region or phylogenetic group. In general, an antivenin neutralizing the venom of one species will be to some degree effective against venoms of closely related species. The crotalid or pit viper antivenin commercially available in the United States is produced against venoms of two species of North American rattlesnakes, the tropical rattlesnake, and one tropical pit viper of the genus *Bothrops*. However, it is clinically effective against venoms of all North American pit vipers and probably most medically important tropical American species. As based on animal experiments, it effectively neutralizes the venoms of some Asian pit vipers but is ineffective against others such as *Agkistrodon acutus, A. rhodostoma, Trimeresurus elegans* and *T. wagleri*. It shows some neutralization of venoms of *Vipera ammodytes* and *V. xanthina palaestinae* but not those of *V. russelii* and *Echis carinatus* (Minton, unpublished). A polyvalent antivenin produced against venoms of several medically important Australian elapid snakes is highly to moderately effective against venoms of a number of Asian, African, and American elapids (Minton, unpublished). On the other hand, venoms of the African mambas show considerable interspecific antigenic differences and are not very effectively neutralized by other mamba antivenins nor by antivenins against other elapid venoms (Christensen and Anderson, 1967).

With the exception of some experimental products prepared in research laboratories, all antivenins are produced by hyperimmunization of horses. A high percentage of individuals who receive antivenin develop serum sickness, and an appreciable number may suffer radiculitis, other neuropathic syndrome, or anaphylaxis (Reid, 1957b; Behar, 1959). The latter complication has resulted in fatalities (Anon., 1957; Trinca, 1963).

For these reasons antivenin should be administered only if there is evidence of systemic or severe local envenomation. A preliminary intradermal or ophthalmic test for hypersensitivity must be done, although a negative reaction does not completely rule out the possibility of anaphylaxis developing.

Because the neutralizing antibody titers of antivenins are much lower than those of most bacterial antitoxins, the dose of antivenin required in a severe snakebite may be very large. Most potentially fatal bites require 30-100 ml. Total doses of 130-170 ml may be needed in severe diamondback rattlesnake bites (McCollough and Gennaro, 1970), and a total of 1150 ml in 19 hours was administered in successful treatment of a king cobra bite (Ganthavorn, 1971). The usual route of administration is by intravenous drip mixed with normal saline to make a 1:10 to 1:50 dilution. Intra-arterial administration is recommended by some (Snyder et al., 1972). Infiltration of antivenin around the area of bite greatly reduces its effectiveness in counteracting systemic symptoms and is no more effective than intravenous administration in preventing or reversing local damage. In the presence of life-threatening systemic envenomation, it is never too late to initiate antivenin therapy, although its effectiveness after 48 hours has not been clearly demonstrated. There is no evidence that antivenin will stop or reverse the late necrotizing effect of either viperid or elapid venoms.

The decision to use or not use antivenin in a given case of snakebite can be difficult for the practicing physician. If signs of severe envenomation are lacking, and the snake is a species known to be not particularly dangerous (e.g., the copperhead in the U.S.), antivenin is not indicated. With a negative test for allergy and no history of previous serum reaction, it is advisable, at least in U.S. practice, to administer antivenin in borderline situations. This is more likely to protect the physician against legal action should an unexpected complication of envenomation develop. Because of the risk of serum sensitization, I tend to be more conservative in recommending the use of antivenin when the patient is a herpetologist or professional snake handler whose occupation subjects him to the risk of repeated bites. If the bite is clearly one that may have a fatal outcome if not actively treated, the use of antivenin is advisable even in the presence of a positive test for horse serum sensitivity. There is seldom time for desensitization, although it may be feasible in some circumstances. If serious serum reactions develop, they can nearly always be reversed if the physician is forewarned. The administration of corticosteroid (e.g. 100 mg hydrocortisone) along with antivenin is often recommended. Antivenin administration has been resumed after the patient has recovered from anaphylaxis (Schapel et al., 1971).

Because antivenins have limited ability to counteract the local necrotizing effects of venoms, other materials have been employed. EDTA (ethylenediamine tetracetic acid) reduces the local necrotizing activity of viperid venoms probably by chelation of divalent metallic ions essential for action of some venom proteases. Prompt local infiltration with EDTA reduced tissue damage in animals injected with cottonmouth venom but had no effect on lethality (Flowers and Goucher, 1965). It was tried clinically against habu bites in the Amami Islands, usually in conjunction with antivenin. Results were inconclusive. Of 192 patients treated, 14 developed severe necrosis and 3 died in contrast to 2 cases of severe necrosis and one death among 26 patients treated only with antivenin. Thiabendazol decreased tissue damage from cottonmouth venom when infiltrated locally in a 0.2 mg/ml/solution (Stone et al., 1966). Metronidazol orally was found helpful in some cases of rattlesnake bite with necrosis (Russell, 1966c). These therapies require further evaluation.

The technique of ligature and cryotherapy has been widely used in the U.S. in an effort to localize the venom in cases of snakebite and limit its damage (Stahnke et al., 1957; Stahnke and McBride, 1966). The results have generally been disappointing, and prolonged use of cold has been attended by severe damage to limbs (McCollough and Gennaro, 1963; Lockhart, 1970; Frank, 1971). Animal experiments have failed to support this mode of therapy (Ya and Perry, 1960; Gill, 1970; Clark, 1971).

Local damage in rattlesnake and other viper bites may be exacerbated by circulatory damage caused by pressure and the direct effect of venom on blood vessels. This is particularly true if venom has been injected beneath the fascia. Fasciotomy or subfascial drainage is recommended in such situations (McCollough and Gennaro, 1970; Weaver, 1970; Lockhart, 1970).

Corticosteroids have been used in snakebite therapy, usually as a supplement to antivenin or other therapeutic agents. Good results were first reported with North American pit viper bites (Wood et al., 1955) and later with both viper and elapid bites in Thailand and India (Benyajati et al., 1960, 1961; Gupta et al., 1960). However, in a well controlled clinical study involving 100 patients bitten by the Malay pit viper, prednisone in a total dose of 120 mg failed to produce improvement in either local or systemic poisoning (Reid et al., 1963c). Trinca (1963) considered corticosteroids of dubious value in poisoning by Australian elapids. Most clinicians in recent years have used corticosteroids chiefly for treatment or prevention of serum reactions or, in doses of 100-500 mg of hydrocortisone or its equivalent, for anti-hypotensive and anti-inflammatory effects. However, repeated large doses of corticosteroids (1000 mg or more of hydrocortisone or its equivalent every 6 hours for 72 hours) have been used in conjunction with fasciotomy as primary treatment for severe rattlesnake bite (Glass, 1969, 1973). Animal experiments have provided no reliable guidelines for use of corticosteroids in snake venom poisoning. Most investigators (Minton, 1952; Schöttler, 1954; Chatterjee, 1969; Clark, 1971) have found them incapable of protecting animals against lethal doses of either viperid or elapid venoms.

Respiratory obstruction from aspiration of oral and respiratory secretions may develop rapidly in elapid envenomation and requires insertion of an endotrachial tube with suction. This may be followed by tracheostomy and use of a respirator. Respiratory failure may respond slowly to antivenin, and artificial support of respiration may be needed for as long as 10 days. Since pharyngeal muscles are also paralyzed, feeding by gastric tube is required (Campbell, 1967, 1969; Pearn, 1971).

Shock in viper bites responds to intravenous fluids with added vasopressor drugs. It may recur unless antivenin is administered in adequate dosage.

Hemodialysis has been used successfully to combat renal failure in sea snake and rattlesnake bites (Sitprija et al., 1971; Danzig and Abels, 1961). In the former instance, it was used as the principal mode of therapy in the absence of suitable antivenin.

Pain is often a prominent part of the picture in viper bites, less so in elapid bites as a rule. Meperidine in doses up to 100 mg may be used if less potent analgesics fail to give relief. Sedatives and tranquilizers in moderate dosage may be given to allay anxiety as may small doses of alcohol. The restlessness and confusion often seen in elapid poisoning result from anoxia and are relieved when this is corrected.

Transfusions are of great value in snakebite when there has been considerable blood loss or hemolysis. Transfusions of blood or plasma from individuals immunized against snake venoms or those who have survived repeated bites have been employed in a few instances. Such donors are rarely available, and the value of their blood in therapy has not been adequately evaluated.

Infection is an occasional complication of snakebite particularly where there has been considerable local tissue damage. Studies of the oral flora of North American pit vipers (Parrish et al., 1956; Fischer et al., 1961; Ledbetter and Kutscher, 1969) and Australian elapids (Williams et al., 1934) have shown the frequent presence of potentially pathogenic *Clostridium* species (but not *C. tetani*) and gram negative bacilli such as *Pseudomonas, Proteus,* and *Entero-bacter.* Organisms may also be introduced by other means such as incisions made in the field and the use of certain folk remedies. Fatal cases of tetanus and gram negative septicemia are known following snakebites. Prophylactic administration of a broad spectrum antibiotic orally and tetanus prophylaxis are advisable in all but obviously trival snakebites. If infection develops, cultures and sensitivity tests guide the choice of antibacterial therapy.

DIAGNOSIS

Differential diagnosis in snakebite involves distinction between snakebite and other injuries, between bites by poisonous and non-poisonous snakes, and between minimal envenomation and severe envenomation when the bite is inflicted by a dangerous species of snake.

Injuries by inanimate objects such as thorns and spines of plants occasionally must be distinguished from snakebite. The nature of the wound and absence of progressive symptoms usually make recognition easy. Wounds made by fish spines and teeth of small mammals are usually deeper and more ragged than those made by snakes' fangs. Wounds inflicted by arthropods such as scorpions, centipedes, spiders and insects are usually smaller than those caused by snakes and do not bleed as freely.

The pattern of puncture wounds will not reliably differentiate bites of poisonous snakes from those of nonpoisonous species. Bites by nonpoisonous snakes may show any pattern from a single puncture to six rows of multiple punctures. Poisonous snake bites showing a single fang puncture are not rare and may terminate fatally. Bites in which the palatine and mandibular teeth cause wounds other than those made by the fangs are also seen. Nonpoisonous snakebites show very little local swelling and cause little pain; unfortunately bites of some dangerous snakes also show little local reaction. Immunological techniques for identification of venom in tissues around the bite have been described (Trethewie and Rawlinson, 1967; Trethewie, 1970). These may permit a reasonably close identification of the snake responsible but are not rapid enough for use in many emergency situations.

Differentiation of minimal from severe envenomation is a matter of clinical judgment. Since symptoms of envenomation may be delayed 12 hours or so in bites by coral snakes and some other elapids, and early symptoms in viper bites may be difficult to in-

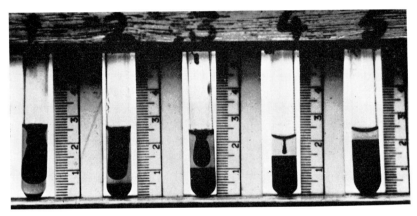

Figure 61. Impairment of clotting is an important sign of systemic envenomation in bites by some species of vipers and can be detected without elaborate equipment. Clot quality after retraction is graded as: (1) normal, (2) slightly defective, (3) moderately defective, (4) severely defective (5) no clot. (Preparation and photo courtesy H. A. Reid.)

terpret at times, it is important that anyone known or strongly suspected of having been bitten by a dangerous snake be kept under close observation, preferably in a hospital, for 24 hours. Development of more than 30 cm of edema and erythema, muscle spasm or fasiculations, hypotension, hemoptysis on coughing, microscopic hematuria, nonclotting blood, and increased prothrombin time are early indications of moderate to severe envenomation by most vipers and those elapids with hemorrhagic venom. In bites by most elapids and those few vipers with strongly neurotoxic venom, ptosis, difficulty in speaking and swallowing, visual disturbances, and drowsiness indicate need for active therapeutic measures. The presence of trismus, muscle pain on passive movement, and myoglobinuria differentiates severe sea snake bites from mild ones and from injuries by other venomous marine organisms.

PROPHYLAXIS AND CONTROL

Attempts to induce active immunity in man by inoculation of minute amounts of snake venom have long been a part of tribal rituals in Africa and elsewhere (Minton and Minton, 1969, pp 89-92). Its effectiveness is extremely dubious. Wiener (1960) reported immunization of a snake handler against tiger snake venom by a series of gradually increasing doses. He eventually attained a fairly high antibody level and about a year after immunization sustained an accidental bite with no evidence of systemic envenomation. Immunization against cobra venom by a course of 17 injections was reported by Flowers (1963). The subject's globulin neutralized 36 mouse LD_{50s} or 0.244 mg of test venom per ml. This basic regimen has been used to immunize a number of high risk individuals in the U.S. One patient whose serum I evaluated about six months after immunization neutralized less than 10 mouse LD_{50}^s per ml or less than 0.05 mg of test venom; the better commercial antivenins neutralize 50-100 LD_{50s} per ml. I know of one immunized individual who sustained an accidental cobra bite. He had no systemic effects but suffered local necrosis.

A large scale trial of immunization with a toxoid prepared by treating habu venom with dihydrothioctic acid was carried out in the Amami and Ryukyu Islands, a region of very high snakebite incidence where 6 to 10 per cent of cases terminate in death or serious disability. A group of 43,446 persons was immunized, and 168 were subsequently bitten by habu snakes. Five suffered necrosis, three with permanent damage, and two died. In a group of 1542 unimmunized persons who were bitten, there were 115 cases of necrosis, 76 with residual disability, and 19 deaths. The results of the immunization were considered encouraging but not conclusive (Sawai et al., 1969). Attempts to produce toxoids of higher immunogenicity against venoms of the habus and some other Far

Eastern snakes are in progress (Sawai and Kawamura, 1969; Kondo et al., 1971).

Control of snakes is best accomplished by indirect measures such as disposal of rubbish that affords them shelter, control of the rodents on which many poisonous snakes feed, and better construction of dwellings. Direct control campaigns have not been successful except in small areas, and may be undesirable as many beneficial snakes are also destroyed. Wearing shoes and clothing that protects the lower legs is advisable where venomous snakes are numerous. There are no effective snake repellents, and any toxicant capable of killing snakes over an extensive area would be dangerous to other forms of life. Introduction of snake predators such as the mongoose is likely to be ineffective or followed by ecological imbalance. However where such predators occur naturally, they should not be disturbed. A poisoning campaign directed against jackals in Israel also decimated the mongoose population and led to an increase in venomous snakes and snakebite (Mendelssohn et al., 1971).

Collecting, keeping, and handling venomous snakes is a risky business. Those who feel it is essential to their professional careers or to fulfilling their inner needs should learn the techniques from responsible individuals and be prepared to follow adequate safety precautions. Zoos and serpentariums should stock antivenins appropriate for those venomous snakes exhibited and have a well rehearsed plan for emergencies.

Chapter 7

VENOMS IN OTHER VERTEBRATES

I. AMPHIBIANS

CONTEMPORARY AMPHIBIANS BELONG to three orders, the widely distributed frogs and toads, the salamanders which are almost entirely confined to the northern hemisphere, and the worm-like caecilians, a small and little known tropical group. For the most part, amphibians are restricted to freshwater and damp terrestrial habitats, although a few, by burrowing and other adaptations, are able to survive in arid lands.

No amphibian is venomous in the sense of being able to inject a toxic substance with the aid of teeth, spines, or other structures. However, many frogs and toads and some salamanders have toxic skin secretions which serve as an effective defense mechanism. The glands producing these secretions may be widely distributed over the body surface or localized in certain areas such as the shoulder region, hind limbs, or tail in salamanders. In many of the toads and some salamanders, the secretion of these glands exudes as a creamy liquid when the animal is injured or seriously threatened.

Toad toxins are for the most part produced in the parotoid glands of the head and shoulder region. The major active components are cardiotoxic steroids known as bufogenins and bufotoxins. Their chemistry and pharmacology has recently been reviewed by Meyer and Linde (1971). Other biologically active constituents include catecholamines such as dopamine and epinephrine and tryptamine bases such as serotonin (Deulofeu and Ruveda, 1971).

Extremely toxic steroidal alkaloids have been isolated from skin secretions of the arrow-poison frogs of the family Dendrobatidae. These are small, brilliantly colored frogs that inhabit rain forest in tropical America. Batrachotoxin obtained from *Phyllobates aurota-enia* has a mouse LD_{50} of about 1 mcg/kg by subcutaneous injec-

179

tion (Daly and Witkop, 1971). Death is accompanied by dyspnea, convulsions, and loss of equilibrium. Its basic activity is one of causing selective and irreversible increase in the permeability of electrogenic membranes to sodium ions. The chemistry and pharmacology have been described in detail by Albuquerque et al. (1971). Similar compounds have been isolated from *Dendrobates pumilio* which exhibits great variability in color and skin toxicity in the Bocas del Toro region of Panama (Daly and Myers, 1967). A dialyzable, non-steroid toxin with mouse LD_{50} of 16 mcg/kg occurs in another group of small, vividly colored tropical American frogs of the family Atelopodidae (Furman et al., 1969). Peptides with cytotoxic and antibiotic activity occur in skin secretions of the unks or fire-bellied toads of Europe and northern Asia (Csordas and Michl, 1969).

Practically all work on the chemistry and pharmacology of salamander skin toxins has been done on the European fire salamander, *Salamandra maculosa*. The skin secretion is a milky, viscid liquid that quickly becomes a gum. The principal toxin is samandarine, an alkaloid with the formula $C_{19}H_{31}NO_2$. The mouse lethal dose is 3.4 mg/kg. Other alkaloids also occur in the secretion. Their chemistry has been reviewed by Habermehl (1971). Samandarine has also been found in the skin of a small Australian frog, *Pseudophryne corroboree* (Habermehl, 1965).

A powerful neurotoxin originally isolated from eggs of California newts *(Taricha* sp.) was found to be identical with tetrodotoxin found in tissues of puffers and certain other fishes of the family Tetraodontidae. The toxin has subsequently been found in the ovary, skin, muscle, and blood of *Taricha* and a related Japanese salamander, *Cynops pyrrhogaster* (Mosher et al., 1964; Kao, 1966).

The toxins of amphibians have little clinical importance. Indirect envenomation by dendrobatid or atelopodid toxins used as arrow poisons probably has occurred, but there is no medically substantiated account. Fatalities have been reported from eating soup containing eggs of the large toad *Bufo marinus* (Licht, 1967). Field herpetologists have reported irritation of the skin and eyes after handling certain tropical American tree frogs. *Phrynohyas spilomma* has been reported to cause violent spells of sneezing and watering of the eyes lasting four to five hours (Duellman, 1956; Lutz, 1971). I have experienced similar although milder symptoms after handling this species and also after collecting spadefoot toads *(Scaphiopus).*

II. LIZARDS

In most sections of the world lizards are the most conspicuous and abundant of reptiles with about 3000 species. Despite a great body of folklore to the contrary, only two of these are venomous. These are the Gila monster and Mexican beaded lizard, the sole living representatives of the family Helodermatidae. They inhabit the southwestern U.S. and western Mexico. Fossils of late Eocene or Oligocene age indicate helodermatids once occurred in western Europe and Colorado. The two living species are large, heavy, rather slow lizards. The Gila monster *(Heloderma suspectum)* is basically a desert animal, but the Mexican beaded lizard *(H. horridum)* inhabits various types of lowland terrain including fairly dense forest. Since they feed exclusively on birds and reptile eggs and nestling birds and mammals, their venom apparently serves a defensive function in deterring counterattacks by females defending their nests.

Figure 62. Gila Monster *(Heloderma suspectum),* the better known of the two species of venomous lizards, occurs in parts of the southwestern U. S. and adjoining Mexico. It is inoffensive and represents a minimal hazard to man. (Photo by Allan Roberts.)

The venom apparatus of these lizards consists of deeply grooved and flanged teeth in the lower jaw with ducts from the venom glands opening near their bases. There are numerous additional ungrooved teeth in both jaws. The venom glands lie under the skin on the outer side of the anterior half of the lower jaw. There are three or four lobes of glandular tissue each with an excretory duct. The secretory cells are cuboidal to columnar and contain secretion granules and vacuoles which discharge their contents into collecting tubules (Phisalix, 1922).

Venom is usually obtained by inducing the lizard to bite a rubber or plastic object and yields of as much as 2 ml have been reported (Arrington, 1930); however, it is very difficult to avoid mixing the venom with other oral secretions. Several observers have noted preliminary teasing of the lizards considerably increases venom secretion. In biting, the lizards hold on tenaciously and occasionally rotate their body on its long axis, a common type of behavior in crocodilians and some nonvenomous lizards. The toxicity and some of the other properties of *Heloderma* venoms are retained for long periods in dried or lyophilized material.

Like snake venoms, lizard venoms contain a mixture of enzymes with at least one nonenzymatic, low molecular weight toxin. Hyaluronidase is present in high concentration, while L-amino acid oxidase, phospholipase fibrinolysin, phosphodiesterase, and protease were found in lower amounts (Styblova and Kornalik, 1967). Serotonin was reported in the venom by Zarafonetis and Kalas (1960). Two arginine esterases, one with kinin-releasing activity, were isolated and separated from the toxic fraction which has not been adequately characterized. A distinctive feature of *Heloderma* venoms is their high resistance to heat. There is no loss of toxicity after heating 20 minutes at 100° in a pH 7 solution, and 70 per cent of the kinin-releasing activity is retained (Mebs and Raudonat, 1966; Mebs, 1972). No significant differences have been found between venoms of the two species of *Heloderma*.

Effects of Heloderma venom have been observed in a variety of animals, and results of earlier experiments have been summarized by Bogert and Martin del Campo (1956). In general, fish, anurans, and reptiles are highly resistant, the lethal doses being in excess of 50 mg per kilo. Gila monsters were found to be highly resistant and unaffected by doses up to 2.25 ml of fresh venom or approximately 45 guinea pig lethal doses. Gila monster serum neutralized about 64 mouse lethal doses of liquid venom per ml, and all of the neutralizing capacity was in the globulin fraction. Heloderma liver extracts also effectively neutralized the venom, but extracts of other organs did not (Tyler, 1956). Experimental determinations of the lethal dose for laboratory mammals have shown great variation that probably reflects variation in the quality of venom preparations used as well as techniques of experimenters. The mouse lethal dose (MLD in some cases; LD_{50} in others) has variously been given as 16 mg per kg (Phisalix, 1922), 10 mg per kg (Cooke and Loeb, 1913), 4 mg per kg (Styblova and Kornalik, 1967), 1.5 mg per kg (Mebs, 1972), and 0.82 mg per kg (Stahnke et al., 1970). The purified toxic fraction had an LD_{50} of 0.43 mg per kg (Mebs, 1972). Rats appear to be considerably more resistant than mice, while toxicity for dogs and rabbits is comparable to that of mice on a body weight basis. Venom of *Heloderm horridum* has been less thoroughly investigated but appears to be somewhat less toxic, and the mouse LD_{50} by subcutaneous injection being 1.4 mg per kg in contrast to 0.82 for H. *suspectum* (Stahnke et al., 1970).

The most striking manifestation of *Heloderma* envenomation in experimental animals reported by earlier investigators was shallow, rapid breathing ultimately with respiratory arrest if fatal doses were given. Weakness, especially in the hind legs, exaggerated reflexes, salivation, vomiting, and increased frequency of urination and defecation were reported. Autopsy findings were not remarkable and included pulmonary congestion of the mucosa (Loeb et al., 1913). In addition to respiratory effects previously reported, intravenous injection of Gila monster venom produced tachycardia, hypotension, and impairment of ventricular contractility (Patterson, 1967a). The venom contained a smooth muscle stimulating factor not blocked by atropine or antihistaminics (Patterson 1967b). Although the venom had no effect on blood coagulation *in vivo* (Patterson and Lee, 1969), massive hemorrhages in the eye, intestine, and kidney have been observed in animals receiving sublethal or marginally lethal doses. The substance responsible for these hemorrhages has not been identified (Mebs, 1972).

Helodermas are unaggressive in their behavior toward man and do not frequent residential or agricultural areas. Virtually all instances of human injury by these lizards have resulted from attempts to capture the animal or from the handling of captive individuals often following considerable provocation of the reptile. As with snakebites, there is a great degree of variability in human response to the bite resulting from the amount of venom injected and probable variation in venom composition and human susceptibility. Woodson (1947) collected 136 case histories of Gila monster bite with 29 fatalities, but there may have been considerable overlap or error in his figures. Bogert and Martin del Campo provided details on 34 cases occurring between 1878 and 1954 eight of which resulted in death. Only one account of a fatal case (Storer, 1931) is accompanied by adequate medical details. In this case the victim's generally poor physical condition may have contributed to the outcome. In two other fatal cases (and in several nonfatal ones) the victim was quite drunk when bitten.

Gila monster bites are quite painful and bleed freely. Local swelling nearly always occurs but is not so extensive nor does it last so long as in rattlesnake or other viperid envenomation. There may be transient lymphangitis. Commonly reported systemic symptoms are tachycardia, tachypnea, nausea, vertigo, and faintness. Less common symptoms include tinnitus, epigastric pain, and a swollen tongue with sore throat. Moderate leukocytosis may be seen. Necrosis does not occur except as a result of secondary infection. A few of the older case reports mentioned prolonged ill health and in one case death a few months after the bite. These may best be ascribed to infection of the bite or to unrelated intercurrent disease. The other fatalities occurred from 52 minutes to a few hours after the bite. In addition to the previous references, case reports have been published by Shannon (1953) and Stahnke et al. (1970). The only well-documented case report of a bite by the Mexican beaded lizard is that of Albritton et al. (1970). The principal complaint was extremely severe pain in the bitten hand and arm accompanied by nausea, vomiting, and moderate swelling. The pulse on admission was 120 and respirations 20. Most of the symptoms subsided after 48 hours.

There is no specific Heloderma antivenin, and I found *H. suspectum* venom totally unreactive in gel diffusion tests against 24 monovalent and polyvalent snake antivenins. Treatment is symptomatic. Meperidine intravenously or xylocaine or other local anaesthetics infiltrated around the bite have been reported to relieve the pain. Tetanus prophylaxis is advisable, and a short course of broad spectrum antibiotic may be administered. There are no guidelines for dealing with severe cardiac or respiratory manifestations of envenomation. Fortunately these seem to be extremely rare.

Diagnosis of Heloderma bite is rarely difficult, since the lizards are large and easily identified. However, Shannon (1954) reported the case of an infant found dead with facial wounds that might have been inflicted by a Gila monster or a rattlesnake. Evidence favored the latter interpretation but was equivocal. Immunological detection of venom in tissue would be of value in such situations. Bites by the larger nonvenomous lizards of the Southwest and Mexico must occasionally be differentiated from those of Helodermas. In the case of the nonvenomous species, the degree of pain and trauma is proportional to the size of the lizard's teeth and the power of its jaws. The very large varanids or monitors of Australia and the Old World tropics can occasionally inflict dangerous wounds; however, these lizards are not venomous.

III. MAMMALS

Few persons associate venom with mammals, yet a well-developed venom apparatus has evolved in two lines of contemporary mammals. The venom apparatus of the platypus *(Ornithorhynchus anatinus)* was described soon after the discovery of this remarkable animal in the rivers of eastern Australia (Hill, 1822; Knox, 1824). This animal and the five species of echidnas or spiny anteaters are the only survivors of the Monotremata or primitive, egg-laying mammals and are restricted today to parts of Australia, Tasmania, and New Guinea. The platypus is semiaquatic and digs long burrows in stream banks from which it emerges to hunt for food in the late afternoon and early morning.

The platypus is unique in that only the male is venomous. A movable horny spur about 15 mm long is located on the inner side of each hind limb near the heel. Venom is produced in a kidney-shaped gland in the inguinal region. The gland consists of numerous alveoli and ducts in a connective tissue stroma and has been reported to undergo a seasonal secretory cycle (Martin and Tidswell, 1895). The venom is conveyed by a long duct to a reservoir at the base of the spur. A second duct leads from the reservoir through a canal in the center of the spur to an opening at its tip.

The fresh venom is a clear opalescent fluid high in protein. At least some activity was retained in a dried sample collected about 30 years previously. This material caused hemorrhagic edema on subcutaneous injection and peripheral vasodilatation with hypotension when injected intravenously. It had weak coagulant and hemolytic activity (Kellaway and Le Messurier, 1935). Fresh material in 20 to 60 mg doses killed rabbits on intravenous injection (Martin and Tidswell, 1895).

The presence of spurs only in males suggests that their function may be chiefly in combat with other males for territory or mates; however, this has not been recorded. There is an account of a captive female seriously wounded by her cagemate of eight years. Calaby (1968) cites several accounts of dogs injured by wounded or cornered platypuses, and I have heard of others from Australian field zoologists. Pain and edema were the chief manifestations of envenomation with dyspnea and weakness in more severe cases. Dogs have been said to die from their injuries.

Human injuries by the platypus were apparently of no great rarity in the last century when the animals were hunted for their fur, but such incidents are now infrequent. The principal manifestations are immediate burning pain which subsides in a few hours but may be followed by soreness for several days. This is accompanied by edema which in some cases has extended from the hand to the shoulder. Lymphadenopathy persisting up to a few weeks is often seen. Faintness, fever, and epigastric pain have been reported. In several cases the wound made by the spur was not followed by any evidence of envenomation (Cleland, 1942; Calaby, 1968).

A spur similar to that of the platypus is seen in male echidnas and small spurs have been reported in females. There is an associated gland and duct, but the presence of venom has not been demonstrated.

The Insectivora, a group of primitive but widely distributed mammals, have a few venomous representatives. Best known of these is the shorttailed shrew *(Blarina brevicauda)*, a very abundant animal in forested parts of the eastern U.S. Its secretive habits permit it to survive in suburbs and city parks. Another well known venomous shrew is the European water shrew *(Neomys fodiens)* which occurs throughout most of Europe and northern Asia. In *Blarina* and probaly in *Neomys* also, venom is produced in specialized cells of the submaxillary glands whose ducts open near the base of the lower incisors. The venom flows along a groove between the elongate median pair of teeth (Pearson, 1950). The principal function of the venom appears to be immobilization of prey. Shrews have a very high metabolic rate and require large amounts of food. Venom facilitates their overcoming comparatively large animals.

Shrew venoms are not well characterized but appear to be proteins. Toxic activity of gland homogenates decreases after a few days at room temperature, but acetone-dried material retains toxicity for years. *Blarina* toxin is stable to heat at 100° for 10 minutes in a neutral or acid medium; however, *Neomys* toxin is destroyed at this temperature. *Blarina* toxin is nondialysable (Ellis and Krayer, 1955; Pucek, 1968).

Crude *Blarina* gland extract had a mouse LD_{50} of 21.8 mg per kg. Toxicity for rabbits was considerably higher. The venom produces hypotension and bradycardia; however, the main effect is on respiration which is initially stimulated but later becomes slower and weaker and finally stops (Ellis and Krayer, 1955; Pearson, 1956). *Neomys* venom produces a profound and irreversible drop in blood pressure as well as respiratory depression; however, its toxicity is less than that of *Blarina* venom (Pucek, 1968).

Shrews are not aggressive toward animals too large to serve as food, but they bite without hesitation if roughly handled. In one reported incident, burning pain and edema promptly followed a *Blarina* bite on the hand. This spread to the forearm and did not resolve completely for two weeks (Maynard, 1889). Others including the author have been bitten by this shrew with nothing more than transient discomfort. Apparently the venom apparatus does not function very effectively in defensive biting.

Evidence of venom has been found in a larger insectivore,

Solenodon paradoxus. The submaxillary glands are large and contain some specialized secretory cells with coarse acidophilic granules. Their ducts end at the base of the second lower incisors which are deeply channeled. Crude submaxillary gland extracts killed mice in comparatively large doses, 380-550 mg per kilo intravenously, while parotid gland extracts more than twice as large were not lethal. The animals showed rapid, gasping respiration, protruding eyes, and convulsions. Captive *Solenodon* have been reported to die after trivial bites inflicted by cage mates (Rabb, 1959).

Solenodon inhabits rock forests in the Dominican Republic, while a related species, *Atopogale cubanus* occurs in Oriente, Cuba. Both are rare, and the Cuban species may actually be extinct. They have a reputation among the natives of being venomous, and a naturalist bitten by *Atopogale* reported inflammation around the punctures made by the lower incisors while those from the upper incisors healed well. Rabb *(op. cit.)* comments, "The utility of the venom for *Solenodon* in its natural environment is unknown and is certainly not indicated by its insectivorus habits. The explanation may be phylogenetic and historical rather than one of present day function."

An anaphylactoid syndrome with circulatory collapse, backache, and hematuria was reported following a bite by the slow loris *(Nycticebus coucang)*, a lemur-like primate found in southeast Asia. There was a good response to epinephrine and an antihistamine. The symptoms were attributed to sensitization to saliva of the loris as a result of previous bites, for another individual bitten by the same animal showed no untoward effects. However, it is perhaps significant that natives of northern Thailand consider the loris venomous, and at least one death has been attributed to its bite (Wilde, 1972).

REFERENCES

Adam, K. R., and Weiss, C.: Scorpion venom. *Zeitschr f Tropenmed u Parasit, 10:*334, 1959.

Adrouny, G. A.; Derbes, V. J., and Jung, R. C.: Isolation of a hemolytic component from fire ant venom. *Science, 130:*449, 1959.

Albritton, D. C.; Parrish, H. M., and Allen, E. R.: Venenation by the Mexican beaded lizard *(Heloderma horridum):* report of a case. *S Dak J Med,* 23:9, 1970.

Albuquerque, E. X.; Daly, J. W., and Witkop, B.: Batrachotoxin: chemistry and pharmacology. *Science, 172:*995, 1971.

Alender, C. B.: A biologically active substance from the spines of two diadematid sea urchins. In Russell, F. E. and Saunders, P. R. (Eds.): *Animal Toxins.* Oxford, Pergamon Press, 1967, pp. 145-155.

Alender, C. B.; Feigen, G. A., and Tomita, J. T.: Isolation and characterization of sea urchin toxin. *Toxicon, 3:*9, 1965.

Aloof-Hirsch, S.; deVries, A., and Berger, A.: The direct lytic factor of cobra venom: purification and chemical characterization. *Biochim Biophys Acta, 154:*53, 1967.

Ameratunga, B.: Middle cerebral occlusion following Russell's viper bite. *J Trop Med Hyg, 75:*95, 1972.

Amorim, M. F., and Mello, R. F.: Intermediate nephron nephrosis from snake poisoning in man. *Am J Pathol, 30:*479, 1954.

Anderson, Philip C.: Treatment of severe Loxoscelism. *Missouri Med, 68:* 609, 1971.

Anderson, S. G., and Ada, G. L.: A lipid component of Murray Valley Encephalitis virus. *Nature, 188:*876, 1960.

Andrews, Edwin H., and Pollard, C. B.: Report of snake bites in Florida and treatment: venoms and antivenoms. *J Fla Med Assn, 40:*388, 1953.

Anon: Medicine and the law. *Lancet 1:*1095, 1957.

Arbesman, C. E.; Langlois, C., and Shulman, S.: The allergic response to stinging insects. *J Allergy, 36:*147, 1965.

Ardao, M. I.; Sosa-Perdomo, C., and Pellaton, M. G.: Venom of the *Megalopyge Urens* (Berg) caterpillar. *Nature, 209:*1139, 1966.

Ariff, A. W.: Cortisone for centipede bites. *Brit Med J, 1:*986, 1956.

Arocha-Piñango, C. L., and Layrisse, M.: Fibrinolysis produced by contact with a caterpillar. *Lancet, 2:*810, 1969.

Arrington, O. N.: Notes on the two poisonous lizards with special reference to *Heloderma suspectum. Bull Antivenin Inst Amer, 4:*32, 1930.

Arvy, L.: Donnees histochimiques sur la glande a venin (glande cheliceri-enne) des araignees dipneumones. *Mem. Inst Butantan Simp Internac, 33:*711, 1966.

Ashley, B. D., and Burchfield, P. M.: Maintenance of a snake colony for the purpose of venom extraction. *Toxicon, 5:*267, 1968.

Atkins, J. A.; Wingo, C. W.; Sodeman, W. A., and Flynn, J. E.: Necrotic arachnidism. *Am J Trop Med Hyg, 7:*165, 1958.

Austin, L.; Carincross, K. D., and McCallum, I. A. N.: Some pharmacological actions of the venom of the stonefish *Synanceja horrida. Arch Inter-nat Pharmacodyn, 131:*339, 1961.

Balozet, Lucien: Venins de scorpions et serum antiscorpionique. *Arch Inst Pasteur Algerie, 33:*90, 1955.

Balozet, Lucien: Le scorpionisme en Afrique du Nord. *Bull Soc Pathol Exot, 57:*33, 1964.

Barnard, J. H.: Severe hidden delayed reactions from insect stings. *N Y State Med J, 66:*1206, 1966.

Barnard, J. H.: Allergic and Pathologic findings in fifty insect sting fatali-ties. *J Allergy, 40:*107, 1967.

Barnes, J. H.: Cause and effect in Irukandji stingings. *Med J Aust, 1:*897, 1964.

Barnes, J. H.: Extraction of cnidarian venom from the living tentacle. In Russell, F. E., and Saunders, P. R. (Eds.) : *Animal Toxins.* Oxford, Per-gamon Press, 1967, pp. 115-129.

Barnes, J. H., and Endean, R.: A dangerous starfish, *Acanthaster planci. Med J Aust, 1:*592, 1964.

Bartholomew, C.: Acute scorpion pancreatitis in Trinidad. *Brit Med J, 1:* 666, 1970.

Baxter, E. H.; Marr, A. G., and Lane, W. R.: Immunity to the venom of the sea wasp *Chironex fleckeri. Toxicon, 6:* 45, 1968.

Baxter, E. H., and Marr, A. G. M.: Sea wasp *(Chironex fleckeri)* venom: lethal, haemolytic and dermonecrotic properties. *Toxicon, 7:*195, 1969.

Beard, R. L.: Insect toxins and venoms. *Ann Rev Entomol, 8:*1, 1963.

Behar, M.: Radiculoneuritis due to antivenom serum treatment after snake bite. *Israel J Med, 18:*21, 1959.

Benton, Allen W.; Morse, R. A., and Stewart, J. D.: Venom collection from honey bees. *Science, 142:*228, 1963.

Benyajati, C.; Koeplung, M., and Sribhibhadh, R.: Viper bite in Thailand with notes on treatment. *J Trop Med Hyg, 63:*257, 1960.

Benyajati, C.; Keoplung, M. and Sribbibhadh, R.: Experimental and clini-cal studies on glucocorticoids in cobra envenomation. *J Trop Med Hyg, 64:*46, 1961.

Bettini, Sergio: Epidemiology of latrodectism. *Toxicon, 2:*93, 1964.

Bhoola, K. D.; Calle, J. D., and Schachter, M.: Identification of acetylcho-line, 5-hydroxytryptamine, histamine, and a new kinin in hornet venom. *J Physiol, 159:*167, 1961.

Birdsey, V.; Lindorfer, J., and Gerwurz, H.: Interaction of toxic venoms with the complement system. *Immunol, 21*:299, 1971.

Birkhead, W. S.: The comparative toxicity of stings of the ictalurid catfish genera *Ictalurus* and *Schilbeodes*. *Comp Biochem Physiol, 22*:101, 1967.

Bitseff, E. L.; Garoni, W. J.; Hardison, C. D., and Thompson, J. M.: The management of stingray injuries of the extremities. *South Med J, 63*:417, 1970.

Bladek, R.: Bee and wasp stings. *Appl Ther, 10*:402, 1968.

Blanquet, Richard: Properties and composition of the nematocyst toxin of the sea anemone, *Aiptasia pallida*. *Comp Biochem Physiol, 25*:893, 1968.

Blanquet, Richard: A toxic protein from the nematocysts of the scyphozoan medusa, *Chrysaora quinquecirrha*. *Toxicon, 10*:103, 1972.

Blum, M. S.: Walker, J. R.; Callahan, P. S., and Novak, A. F.: Chemical, insecticidal, and antibiotic properties of fire ant venom. *Science, 128*:306, 1958.

Bogert, Charles M., and Martin del Campo, R.: The Gila Monster and its allies. *Bull Am Mus Nat Hist, 109*:1-238, 1956.

Bonilla, C. A., and Fiero, M. K.: Comparative biochemistry and pharmacology of salivary gland secretions. II. Chromatographic separation of the basic proteins from some North American rattlesnake venoms. *J Chromatogr, 56*:253, 1971.

Bonilla, C. A.; Fiero, M. K., and Frank, L. P.: Isolation of a basic protein neurotoxin from *Crotalus adamanteus* venom. In DeVries, A., and Kochva, E.: *Toxins of Animal and Plant Origin,* New York, Gordon and Breach, 1971, vol. 1, p. 343.

Bonnett, D. E., and Guttman, S. I.: Inhibition of moccasin *(Agkistrodon piscivorus)* venom proteolytic activity by the serum of the Florida king snake *(Lampropeltis getulus floridana. Toxicon, 9*:417, 1971.

Boquet, P.: Sur les properties antivenimeuses du serum de *Vipera aspis. Ann Inst Pasteur, 71*:340, 1945.

Boquet, P.: Venins de serpents *(lere partie)* physio-pathologie de l'envenomation et proprities biologiques des venins. *Toxicon, 2*:5, 1964.

Boquet, P.; Izard, Y.; Jouannet, M., and Meaume, J.: Enzymes et toxines des venins de serpents. Recherches biochemiques et immunologiques sur le venin de *Naja nigricollis. Mem Inst Butantan Simp Internac, 33*:371, 1966.

Botes, D. P.: Snake venom toxins. The amino acid sequence of toxins b and d from *Naja melanoleuca* venom. *J Biol Chem, 247*:2866, 1972.

Botes, D. P., and Strydom, D. J.: A neurotoxin from Egyptian cobra *(Naja haje haje)* venom. *J Biol Chem, 244*:4147, 1969.

Botes, D. P.; Strydom, D. J.; Anderson, C. G., and Christensen, P. A.: Snake venom toxins. Purification and properties of three toxins from *Naja nivea* (Linnaeus) (Cape cobra) venom. *J Biol Chem, 246*:3132, 1971.

Bouisset, L. and Larrouy, G.: Envenimations par *Scorpio maurus* et *Buthus occitanus* dans le Department de Tlemcen. *Bull Soc Pathol Exot, 55:*139, 1962.

Brade, V., and Vogt, W.: Immunization against cobra venom. *Experentia, 27:*1338, 1971.

Braganca, B. M., and Khandeparkar, V. G.: Phospholipase C activity of cobra venom and lysis of Yoshida sarcoma cells. *Life Sci, 5:*1911, 1966.

Braganca, B. M., and Sambray, Y. M.: Multiple forms of cobra venom phospholipase A. *Nature, 216:*1210, 1967.

Brand, J. M.; Blum, M. S.; Fales, H. M., and MacConnell, J. G.: Fire ant venoms: comparative analyses of alkaloidal components. *Toxicon, 10:* 259, 1972.

Breder, C. M.: Defensive behavior and venom in *Scorpaena* and *Dactylopterus. Copeia,* no. 4 pp. 698, 1963.

Bristowe, W. S.: *A Book of Spiders.* London, King Penguin, 1947.

Brown, L. L.: Fire ant allergy. *South Med J, 65:*273, 1972.

Browne, S. G.: Cantharidin poisoning due to a "blister beetle." *Br Med J, 2:*1290, 1960.

Bücherl, Wolfgang: Acco do veneno dos escolopendromorfos do Brasil sobre alguns animais de laboratorio. *Mem Inst Butantan, 19:*181, 1946.

Bücherl, W.: Escorpiones e escorionismo no Brasil. *Mem Inst Butantan, 25:* 53, 1953.

Bücherl, W.: Über die Ermittlung von Durchschnitt und Höchst-Giftmengen bei den haüfigsten Giftschlangen Sudamerikas. In Behringwerke Mitteilungen *Die Giftschlangen der Erde.* Marburg-Lahn, N. G. Elwert, 1963, pp. 67-120.

Bücherl, W.: Biologia de artropodos peconhentos. *Mem Inst Butantan, 31:* 85, 1964.

Burnett, H. J. W.; Stone, J. H.; Pierce, L. H.; Cargo, D. G.; Layne, E. C., and Sutton, J. S.: A physical and chemical study of sea nettle nematocysts and their toxin. *J Invest Derm, 51:*330, 1968.

Bursoum, G. S.; Nawaby, M., and Salama, S.: Scorpion poisoning, its signs, symptoms and treatment. *J Egypt Med Assn, 37:*857, 1954.

Burtt, E.: Exudate from millipedes with particular reference to its injurious effects. *Trop Dis Bull, 44:*7, 1947.

Cadzow, W. H.: Puncture wound of the liver by stingray spines. *Med J Aust, 1:*936, 1960.

Calaby, J. H.: The platypus *(Ornithorhynchus anatinus)* and its venomous characteristics. In Bücherl, W.; Buckley, E. E., and Deulofeu, V.: *Venomous Animals and their Venoms.* New York, Academic Press, 1968, vol. 1, pp. 15-29.

Cameron, A. M., and Endean, R.: The venom glands of teleost fishes. *Toxicon, 10:*301, 1972.

Campbell, C. H.: Venomous snake bite in Papua and its treatment. *Trans Roy Soc Trop Med Hyg, 58:*263, 1967.

Campbell, C. H.: Clinical aspects of snake bite in the Pacific area. *Toxicon,* 7:25, 1969.

Campbell, E. G.: Tick paralysis. *J Kan Med Soc, 65*:465, 1964.

Caro, M. R.; Verbes, V. J., and Jung, R.: Sting responses to the sting of the imported fire ant *(Solenopsis saevissima). Arch Derm, 75*:475, 1957.

Castex, M.: Clinica y therapeutica de la enfermedad paratrygonica. *Rev Asoc Med Argent, 79*:547, 1965.

Castex, M.: Fresh water venomous rays. In Russell, F. E., and Saunders, P. R. (Eds.): *Animal Toxins.* Oxford, Pergamon Press, 1967, pp. 167-176.

Cavill, G. W. K.; Robertson, P., and Whitfield, F. B.: Venom and venom apparatus of the bull ant. *Myrmecia gulosa* (Fabr.). *Science, 79*:80, 1964.

Cavill, G. W. K., and Robertson, Phyllis: Ant venoms, attractants and repellants. *Science, 153*:1137, 1965.

Chang, C. C., and Lee, C. Y.: Isolation of neurotoxins from the venom of *Bungarus multicinctus* and their modes of neuromuscular blocking action. *Arch Int Pharmacodyn, 144*:241, 1963.

Chapman, D. S.: The symptomatology, pathology, and treatment of the bites of venomous snakes of Central and Southern Africa. In Bücherl, W.; Buckley, E. E., and Deulofeu, V.: *Venomous Animals and their Venoms.* New York, Academic Press, 1968, vol. 1, pp. 467-476.

Chatterjee, S. C.: An experimental study on hydrocortisone in cobra envenomation. *J Indian Med Assn, 52*:493, 1969.

Cherington, M., and Snyder, R. D.: Tick paralysis neurophysiologic studies. *New Eng J Med, 278*:95, 1968.

Cheymol, J.; Bourillet, F., and Roch, M.: Action neuromusculaire des venins de quelques Crotalidae, Elapidae et Hydrophiidae. *Mem Inst Butantan Simp Internac, 33*:541, 1966.

Chowhan, J. S.: Emergencies in snake-bite poisoning and how to treat them. *The Antiseptic, 35*:544, 1938.

Christensen, Poul A.: South African Snake Venoms and Antivenoms. The South African Institute for Medical Research, Johannesburg, 1955.

Christensen, P. A.: The venoms of central and south African snakes. In Bücherl, W.; Buckley, E. E., and Deulofeu, V.: *Venomous Animals and their Venoms.* New York, Academic Press, 1968, vol. 1, p. 437.

Christensen, P. A.: The treatment of snakebite. *S Afr Med J, 43*:1253, 1969.

Christensen, P. A., and Anderson, C. G.: Observations on *Dendroaspis* venoms. In Russell, F. E., and Saunders, P. R.: *Animal Toxins.* Oxford, Pergamon Press, 1967, p. 223.

Clark, R. W.: Cryotherapy and corticosteroids in the treatment of rattlesnake bite. *Milit Med, 136*:42, 1971.

Clark, W. C., and Voris, H. K.: Venom neutralization by rattlesnake serum albumin. *Science, 164*:1302, 1969.

Cleland, J. B.: Injuries and diseases in Australia attributable to animals (insects excepted). *Med J Aust, 2*:313, 1942.

Cleland, John B., and Southcott, R. V.: *Injuries to Man from Marine Invertebrates in the Australian Region.* Canberra, Commonwealth of Australia National Health and Medical Research Council, 1965.

Cochrane, C. G.; Muller-Eberhard, H. J., and Aikin, B. S.: Depletion of plasma complement in vivo by a protein of cobra venom: its effect on various immunological reactions. *J Immunol, 105:55,* 1970.

Cohen, I.; Margalith, Z.; Kaminsky, E., and deVries, A.: Isolation and characterization of kinin-releasing enzyme of *Echis coloratus* venom. *Toxicon, 7:3,* 1969.

Condrea, E., and deVries, A.: Venom phospholipase A: A review. *Toxcon, 2:261,* 1965.

Corkey, J. A.: Ophthalmia nodosa due to caterpillar hairs. *Br J Ophthal, 39:301,* 1955.

Corkill, N. L.; Ionides, C. J. P., and Pitman, C. R. S.: Biting and poisoning by the mole vipers of the genus *Atractaspis. Trans Roy Soc Trop Med Hyg, 53:95,* 1959.

Corson, E. F., and Pratt, A. G.: "Red moss" dermatitis. *Arch Derm Syph, 47:* 574, 1943.

Csordas, A., and Michl, H.: Primary structure of two oligopeptides of the toxin of *Bombina variegata. Toxicon, 7:103,* 1969.

D'Ajello, V.; Mauro, A., and Bettini, S.: Effect of the venom of the black widow spider *Latrodectus tredecimguttatus* on evoked action potential in the isolated nerve cord of the *Periplaneta americana. Toxicon, 7:139,* 1969.

D'Ajello, V.; Magni, F., and Bettini, S.: The effect of the venom of the black widow spider *Latrodectus mactans tredecimguttatus* on the giant neurones of *Periplaneta americana. Toxicon, 9:103,* 1971.

Daly, J. W., and Myers, C. W.: Toxicity of Panamanian poison frogs *(Dendrobates)* some biological and chemical aspects. *Science, 156:970,* 1967.

Daly, J. W., and Witkop, B.: Chemistry and pharmacology of frog venoms. In Bücherl, W.; Buckley, E. E., and Deulofeu, V.: *Venomous Animals and their Venoms.* New York, Academic Press, 1968, vol II, pp. 497-519.

Damus, P. S.; Markland, F. S.; Davidson, T. M., and Shanley, J. D.: A purified procoagulant enzyme from the venom of the eastern diamondback rattlesnake (Crotalus adamanteus): *in vivo* and *in vitro* studies. *J Lab Clin Med, 79:906,* 1972.

Danzig, L. E., and Abels, G.: Hemodialysis of acute renal failure following rattlesnake bite. *J A M A, 175:136,* 1961.

DaSilva, T. L.: The scorpion problem in Ribeirao Preto. São Paulo, Brazil. Notes on epidemiology and prophylaxis. *Am J Trop Med Hyg, 1:508,* 1952.

Davenport, R. C., and Budden, F. H.: Loss of sight following snake bite. *Br J Ophthal, 37:119,* 1953.

Davidson, R. A.: Case of African cobra bite. *Br Med J, 4:660,* 1970.

Day, J. M.: Death due to cerebral infarction after wasp stings. *Arch Neurol,* 7:184, 1962.

Deakins, D. E., and Saunders, P. R.: Purification of the lethal fraction of the venom of the stonefish *Synanceja horrida* (Linnaeus). *Toxicon, 4:257,* 1967.

Delgado, A.: Investigacion ecologica sobre *Loxosceles rufipes* (Lucas) en la region costerax del Peru. *Mem Inst Butantan Simp Internac, 33:*683, 1966.

Denny, W. F.; Dillaha, C. J., and Morgan, P. N.: Hemotoxic effect of *Lox. reclusus venom. In vivo* and *in vitro* studies. *Jour Lab Clin Med, 64:*291, 1964.

Denson, K. W. E.: Coagulant and anticoagulant action of snake venoms. *Toxicon, 7:*5, 1969.

Deoras, P. J.: Studies on Bombay snakes: snake farm yield records and their probable significance. In Keegan, H. L., and Macfarlane, W. V.: *Venomous and Poisonous Animals and Noxious plants of the Pacific Region.* Oxford, Pergamon Press, 1963, pp. 337-349.

Deoras, P. J., and Vad, N. E.: The milking of scorpions. *Toxicon, 1:*41, 1962.

DeOreo, G. A.: Dermatitis venenata resulting from contact with marine animals (hydroids). *Arch Derm Syph, 54:*637, 1946.

Deulofeu, V., and Ruveda, E. A.: The basic constituents of toad venoms. In Bücherl, W.; Buckley, E. E., and Deulofeu, V.: *Venomous Animals and their Venoms,* 1971, vol. II, pp. 475-495.

Devi, C. S.; Reddy, C. N.; Devi, S. L.; Subrahmanyam, Y. R.; Bhatt, H. V.; Suvarnakumari, G.; Murthy, D. P., and Reddy, C. R. R. M.: Defibrination syndrome due to scorpion venom poisoning. *Br Med J, 1:*345, 1970.

deVries, A.; Kirschman, C,; Klibansky, C.; Condrea, E., and Gitter, S.: Hemolytic action of indirect lytic snake venom *in vivo. Toxicon, 1:*19, 1962.

Dillaha, C. J.; Jansen, G. T.; Honeycutt, W. M. and Hayden, C. R.: North American loxoscelism. *JAMA, 188:*33, 1964.

Diniz, C. R.: Bradykinin formation by snake venom. In Bücherl, W.; Buckley, E. E., and Deulofeu, V.: *Venomous Animals and their Venoms.* New York, Academic Press, 1968, vol. I, pp. 217-228.

Doery, H. M., and Pearson, J. E.: Haemolysins in venoms of Australian snakes. *Biochem J, 78:*820, 1961.

Doery, H. M., and Pearson, J. E.: Phospholipase B in snake venoms and bee venom. *Biochem J, 92:*599, 1964.

Dreyer, V.: Ophthalmia nodosa. *Acta Ophthal, 31:*429, 1953.

Duellman, William E.: The frogs of the hylid genus *Phrynohyas* Fitzinger, *Misc Pub Mus Zool Univ Mich,* 1956.

Duran-Reynals, F.: Invasion of the body by animal poisons. *Science, 83:*286, 1936.

Eads, R. B.; Menzies, G. C., and Hightower, B. G.: The ticks of Texas with notes on their medical significance. *Tex J Sci, 55:*7, 1956.

Eaker, D.: Isolation of neurotoxin in a case of fatal cobra bite. *J Forensic*

*Med, 16:*96, 1969.

Edery, H.; Ishay, J.; Lass, I., and Gitter, S.: Pharmacological activity of oriental hornet *(Vespa orientalis)* venom. *Toxicon, 10:*13, 1972.

Edmonds, C.: A non-fatal case of blue-ringed octopus bite. *Med J Aust, 2:* 601, 1969.

Edwards, J. S.: The action and composition of the saliva of an assassin bug *Platymeris rhadamanthus* Gaerst (Hemiptera, Reduviidae). *J Exptl Biol, 38:*61, 1961.

Efrati, P.: Bites by *Loxosceles* spiders in Israel. *Toxicon, 6:*239, 1969.

Efrati, P., and Reif, L.: Clinical and pathological observations on sixty-five cases of viper bite in Israel. *Am J Trop Med Hyg, 2:*1085, 1953.

Eisner, T., and Meinwald, J.: Defensive secretions of arthropods. *Science, 153:*1341, 1966.

Ellis, E. F., and Smith, R. T.: Systemic anaphylaxis after rattlesnake bite. *JAMA, 193:*401, 1965.

Ellis, S., and Krayer, O.: Properties of a toxin from the salivary gland of the shrew *(Blarina brevicauda). J Pharmacol Exptl Therap, 114:*127, 1955.

Endean, R., and Rudkin, Clare: Studies on the venoms of some Conidae. *Toxicon, 1:*49, 1963.

Endean, R., and Rudkin, C.: Further studies on the venoms of Conidae. *Toxicon, 2:*225, 1965.

Endean, R., and Duchemin, C.: The venom apparatus of *Conus magus. Toxicon, 4:*275, 1967.

Endean, R.; Duchemin, C.; McColm, D., and Fraser, E. H.: A study of the biological activity of the toxic material derived from nematocysts of the cubomedusan *Chironex fleckeri. Toxicon, 6:*179, 1969.

Endean, R., and Noble, Mary: Toxic material from the tentacles of the cubomedusan *Chironex fleckeri. Toxicon, 9:*255, 1971.

Erspamer, V., and Anastasi, A.: Structure and pharmacological actions of eledoisin, the active endecapeptide of the posterior salivary glands of *Eledone. Experientia, 18:*58, 1962.

Esnouf, M. P., and Tunnah, G. W.: The isolation and properties of the thrombin-like activity from *Ancistrodon rhodostoma* venom. *Br J Haematol, 13:*581, 1967.

Espinoza, N. C.: Accion del veneno de *Haplopus limensis. Mem Inst Butantan Simp Internac, 33:*799, 1966.

Ewing, H. E.: Observations on the habits and the injury caused by the bites and stings of some common North American arthropods. *Am J Trop Med, 8:*39, 1928.

Fain, A.: Toxic action of rove beetles *(Coleoptera Staphylinidae). Mem Inst Butantan Simp Internac, 33:*835, 1966.

Feigen, George A.; Sanz, E., and Alender, C. B.: Studies on the mode of action of sea urchin toxin I. Conditions affecting release of histamine and other agents from isolated tissues. *Toxicon, 4:*161, 1966.

Fiero, M. K.; Seifert, M. W.; Weaver, T. J., and Bonilla, C. A.: Comparative study of juvenile and adult prairie rattlesnake *(Crotalus viridis viridis)* venoms. *Toxicon, 10:*81, 1972.

Finlayson, M. H.: Spider-bite in South Africa. *S Afr Med J, 29:*509, 1955.

Fischer, F. G., and Bohn, Hans: Die Giftsekrete der brasilianischen Tarantel *Lycosa erythrognatha* und der Wanderspinne *Phoneutria fera. Zeit f Physiol. Chem, 306:*265, 1957.

Fischer, F. J.; Ramsey, H. W.; Simon, J., and Gennaro, J. F.: Antivenin and antitoxin in the treatment of experimental rattlesnake venom intoxication *(Crotalus adamanteus). Am J Trop Med Hyg, 10:*75, 1961.

Fish, C. J., and Cobb, M. C.: Noxious marine animals of the central and western Pacific Ocean. *U. S. Fish Wildlife Serv Rpt No. 36,* pp. 17-20; 1954.

Flecker, H.: Cone shell mollusc poisoning with report of a fatal case. *Med J Aust, 1:*464, 1936.

Flecker, H.: Irukandji sting to north Queensland bathers without production of weals but with severe general symptoms. *Med J Aust, 2:*89, 1952.

Flecker, H., and Cotton, B. C.: Fatal bite from octopus. *Med J Aust, 2:*329, 1955.

Flowers, H. H.: Active immunization of a human being against cobra *(Naja naja)* venom. *Nature, 200:*1017, 1963.

Flowers, H., and Goucher, C.: The effect of EDTA on the extent of tissue damage caused by the venoms of *Bothrops atrox* and *Agikistrodon piscivorus. Toxicon, 2:*221, 1965.

Foot, N. C.: Pathology of the dermatitis caused by *Megalopyge opercularis,* a Texan caterpillar, *J Exptl Med, 35:*737, 1922.

Foubert, E. L., and Stier, R. A.: Antigenic relationships between honeybees, wasps, yellow hornets, black hornets, and yellowjackets. *J Allergy, 29:*13, 1958.

Frank, H. A.: Snakebite or frostbite: what are we doing? *Calif Med, 114:*25, 1971.

Frank, L.: Black widow spider bite syndrome. *Mil Surg, 91:*329, 1942.

Frazier, C. A.: Allergic reactions to insect stings: a review of 180 cases. *South Med J, 57:*1028, 1964.

Fredholm, B., and Haegermark, Ö.: Histamine release from rat mast cells induced by a mast cell degranulating fraction in bee venom. *Acta Physiol. Scand, 69:*304, 1967.

Fredholm, B.; Haegermark, Ö., and Strandberg, K.: Release of histamine and formation of smooth-muscle stimulating principles in guinea pig lung tissue induced by antigen and bee venom phosphitidase A. *Acta Physiol Scand, 76:*446, 1969.

Freeman, S. E., and Turner, R. J.: A pharmacological study of the toxin of a Cnidarian *Chironex fleckeri* Southcott. *Brit J Pharmacol, 35:*510, 1969.

Freyvogel, T. A.; Honegger, C. G., and Maretic, Z.: Zur Biologie und Giftigheit der ostafrikanischen Vogelspinne *Pterinochilus*. *Acta Trop, 25*:217, 1968.

Friederich, C., and Tu, A.: Role of metals in snake venoms for hemorrhagic, esterase, and proteolytic activities. *Biochem Pharmacol, 20*:1549, 1971.

Froes, H. P.: Studies on venomous fishes of tropical countries. *J Trop Med Hyg, 36*:134, 1933.

Frontali, N., and Grasso, A.: Separation of three toxicologically different protein components from the venom of the spider *Latrodectus tredecimguttatus*. *Arch Biochem Biophys, 106*:213, 1964.

Fuhrman, F. A.; Fuhrman, G. J., and Mosher, H. S.: Toxin from skin of frogs of the genus *Atelopus:* differentiation from dendrobatid toxins. *Science, 165*:1376, 1969.

Furman, D. P., and Reeves, W. C.: Toxic bite of a spider *Chiracanthium inclusum. Calif Med, 87*:114, 1957.

Furtado, M. A., and Lester, I. A.: Myoglobinuria following snakebite. *Med J Aust, 1*:674, 1968.

Fujiwara, T.: On the poisonous pedicellaria of *Toxopneustes pileolus* (Lamarch). *Annot Zool Japon, 15*:62, 1935.

Gajardo-Tobar, R.: El araneismo en el mundo tropical y subtropical. *Mem Inst Butantan Simp Internac, 33*:45, 1966. Mi experiencia sobre Loxoscelismo. *Ibid.*, pp. 689-698.

Gans, Carl, and Elliott, W. B.: Snake venoms: production, injection, action. *Adv Oral Biol, 3*:45, 1968.

Ganthavorn, S.: A case of king cobra bite. *Toxicon, 9*:293, 1971.

Garriott, James C., and Lane, C. E.: Some autonomic effects of *Physalia* toxin. *Toxicon, 6*:281, 1969.

Gennaro, J. F.: Observations on the treatment of snakebite in North America. In Keegan, H. L., and Macfarlane, W. V.: *Venomous and Poisonous Animals and Noxious Plants of the Pacific Region*. Oxford, Pergamon Press, 1963, pp. 427-446.

Gennaro, J. F.; Squicciarina, P. J.; Heisler, M., and Hall, H. P.: The microscopic anatomy and histochemistry of the poison apparatus of the cottonmouth moccasin *(Ancistrodon p. piscivorus). Anat Rec, 136*:196, 1960.

Gennaro, J. F.; Leopold, R. S., and Merriam, T. W.: Observations on the actual quantity of venom introduced by several species of crotalid snakes in their bite. *Anat Rec, 139*:303, 1961.

Gennaro, J. F.; Anton, A. H., and Sayre, D. F.: The fine structure of pit viper venom and additional observation on the role of aromatic amines in the physiology of the pit viper. *Comp Biochem Physiol, 25*:285, 1968.

Ghiretti, F.: Cephalotoxin: the crab-paralyzing agent of the posterior salivary glands of cephalopods. *Nature, 183*:1192, 1959.

Ghosh, B.; De, S. S., and Chowdhuri, D. K.: Separation of the neurotoxin from the crude cobra venom and study of the action of a number of reducing agents on it. *Indian J Med, Res, 29:*367, 1941.

Gilkes, M. J.: Snake venom conjunctivitis. *Br J Ophthal, 43:*638, 1959.

Gill, K. A. Jr.: The evaluation of cryotherapy in the treatment of snake envenomization. *Southern Med J, 63:*552, 1970.

Ginsberg, N. J.; Dauer, M., and Slotta, K.: Melittin used as a protective agent against x-irradiation. *Nature, 220:*1334, 1968.

Gitter, S.; Kochwa, S.; deVries, A.; Levi, G.; Rechnic, J., and Caspar, J.: Studies on snake venoms of the Near East. *Am J Trop Med Hyg, 11:*861, 1962.

Glass, T. G. Jr.: Cortisone and immediate fasciotomy in the treatment of severe rattlesnake bite. *Texas Med, 65:*40, 1969.

Glass, T. G. Jr.: Early debridement in pit viper bite. *Surg Gynecol Obstet, 136:*774, 1973.

Goldman, L.; Sawyer, F.; Levine, A.; Goldman, J.; Goldman, S., and Springer, J.: Investigative studies of skin irritations from caterpillars. *J Invest Derm, 34:*67, 1960.

Goncalves, J. M.: Purification and properties of crotamine. In Buckley, E. E., and Porges, N. (Eds.): *Venoms,* AAAS Publication 44. American Association for Advancement of Science, Washington, D. C., 1956, pp. 261-274.

Goncalves, J. M., and Giglio, J. R.: Amino-acid composition and terminal group analysis of crotamine. *Proc 6th Int Congr Biochem N Y, 2:*170, 1964.

Gorham, J. R.: The brown recluse spider and necrotic spiderbite. *J Environ Health, 31:*1, 1968.

Gorham, J. R., and Rheney, T. B.: Envenomation by the spiders *Chiracanthium inclusum* and *Argiope aurantia. J A M A, 206:*1958, 1968.

Grasset, E.; Zoutendyk, A., and Schaafsma, A. W.: Studies on the toxic and antigenic properties of South African snake venoms with special reference to the polyvalency of South African antivenene. *Trans Roy Soc Trop Med Hyg, 28:*601, 1935.

Grasset, E.; Schaafsma, A., and Hodgson, J. A.: Studies on the venom of South African scorpions *(Parabuthus, Hadogenes, Opisthophthalmus)* and the preparation of a specific anti-scorpion serum. *Trans Roy Soc Trop Med Hyg, 39:*397, 1946.

Greer, William: Arachnidism. *New Eng J Med, 240:*5, 1949.

Grothaus, R. H., and Teller, L. W.: Envenomation by the spider *Lycosa miami* Wallace. *J Med Ent, 5:*500, 1968.

Gudger, E. W.: Is the stingray's sting poisonous? *Bull Hist Med, 14:*467, 1943.

Gueron, M., and Yarom, R.: Cardiovascular manifestations of severe scorpion sting. *Chest, 57:*156, 1970.

Gundersen, T.; Heath, P., and Garron, L. K.: Ophthalmia nodosa. *Trans Am Ophthal Soc, 48:*151, 1950.

Gupta, P. S.; Bhargava, S. P., and Sharma, M. L.: A review of 200 cases of snake bite with special reference to the corticosteroid therapy. *J Indian Med Assn, 35*:387, 1960.

Guttman-Friedman, A.: Blindness after snake-bite. *Br J Ophthal, 40*:57, 1956.

Haarvaldsen, R., and Fonnum, F.: Weever venom. *Nature, 199*:286, 1963.

Habermann, E.: Bee and wasp venoms. *Science, 177*:314, 1972.

Habermann, E., and Reiz, K. G.: Biochemie der Bienengiftpeptide Melittin und Apamin. *Biochem Zeit, 343*:192, 1965.

Habermann, E., and Rubsamen, K.: Biochemical and pharmacological analysis of the so-called crotoxin. In DeVries, A., and Kochva, E.: *Toxins of Animal and Plant Origin*. New York, Gordon and Breach, 1971, vol. I, p. 333.

Habermehl, Gerhard: Uber eine alkaloidartige Substanz aus der Haut von *Pseudophryne corroboree*. *Zeit Naturforsch, 20*:1129, 1965.

Habermehl, G.: Toxicology, pharmacology, chemistry, and biochemistry of salamander venom. In Bücherl, W.; Buckley, E. E., and Deulofeu, V.: *Venomous Animals and their Venoms*. 1971, vol. II, pp. 559-584.

Hadar, H., and Gitter, S.: The results of treatment with Pasteur antiserum in cases of snake bite. *Harafuah, 56*:1, 1959.

Hadler, W. A., and Vital Brazil, O.: Pharmacology of crystalline crotoxin. IV. Nephrotoxicity. *Mem Inst Butantan Simp Internac, 33*:1001, 1966.

Hall, M. C.: Lesions due to the bite of the wheel-bug, *Arilus cristatus* (Hemiptera; Reduviidae). *Arch Int Med, 33*:513, 1924.

Halstead, B. W.: Weever stings and their medical management. *U. S. Armed Forces Med J, 8*:1441, 1957.

Halstead, Bruce W.: *Poisonous and Venomous Marine Animals of the World*. Washington, D. C., U. S. Government Printing Office, vol. I, 1965; vol III, 1970.

Halstead, B. W., and Ryckman, R.: Injurious effects from contact with millipedes. *Med Arts Sci, 3*:16, 1949.

Halstead, B. W., and Bunker, N. C.: The venom apparatus of the ratfish, *Hydrolagus colliei*. *Copeia*, no. 2, 128, 1952.

Halstead, B. W., and Bunker, N. C.: Stingray attacks and their treatment. *Am J Trop Med Hyg, 2*:115, 1953.

Halstead, B. W.; Kuninobu, L. S., and Hebard, H. G.: Catfish stings and the venom apparatus of the Mexican catfish *Galeichthys felis* (Linnaeus). *Trans Amer Microscop Soc, 72*:297, 1953.

Halstead, B. W.; Ocampo, R. R., and Modglin, F. R.: A study on the comparative anatomy of the venom apparatus of certain North American stingrays. *J Morph, 97*:1, 1955.

Halstead, B. W.; Danielson, D. D.; Baldwin, W. J., and Engen, P. C.: Morphology of the venom apparatus of the leather-back fish *Scomberoides sanctipetri* (Cuvier). *Toxicon, 10*:249, 1972.

Haneveld, G. T.: Beten door reuzenduizendpoten van Niew-Guinea *(Scolo-*

pendra morsitans en *Sc. subspinipes*). *Nederl Tijdschr v Geneesk, 100:* 2906, 1956.

Haneveld, G. T.: Eye lesions caused by the exudate of tropical millipedes. I. Report of a case. *Trop Geograph Med, 10:*165, 1958.

Henderson, F. W.: Tick paralysis. *J A M A, 175:*615, 1961.

Hercus, J. P.: An unusual eye condition. *Med J Aust, 1:*98, 1944.

Herre, A.W.C.T.: A case of poisoning by a stinging catfish in the Philippines. *Copeia,* no. 3, p. 222; 1949.

Hessinger, D. A., Lenhoff, H. M. and Kahan, L.: Hemolytic phospholipase A and nerve affecting activities of se aanemone nematocyst venom. *Nature (New Biol), 241:*125, 1973.

Hill, P.: On the *Ornithorhynchus paradoxus;* its venomous spur and general structure. *Trans Lin Soc, 13:*622, 1822.

Hill, W. R.; Rubenstine, A. D., and Kovacs, J.: Dermatitis resulting from contact with moths (genus *Hylesia*). *J A M A, 138:*737, 1948.

Högberg, B., and Uvnas, B.: Further observations on the disruption of rat mesentery mast cells caused by compound 48/40, antigen-antibody reactions, lecithinase A, and decylamine. *Acta Physiol Scand, 48:*133, 1960.

Homma, M., and Tu, A. T.: Morphology of local tissue damage in experimental snake envenomation. *Br J Exp Pathol, 52:*538, 1971.

Hurwitz, B. J., and Hull, P. R.: Berg-adder bite. *S Afr Med J, 45:*969, 1971.

Ibrahim, S. A.: A study on sea-snake venom phospholipase A. *Toxicon, 8:* 221, 1970.

Ingram, W. W., and Musgrave, A.: Spider bite (arachnidism): a survey of its occurrence in Australia. *Med J Aust, 2:*10, 1933.

Ishay, J.: Gitter, S., and Fischel, J.: The production and effective of rabbit antiserum against *Vespa orientalis* venom. *Acta Allergol, 26:*286, 1971.

James, M. T., and Harwood, R. F.: *Herm's Medical Entomology,* 6th ed., London, Macmillan, 1969, pp. 327-341.

Jaques, R.: The hyaluronidase content of animal venoms. In Buckley, E. E., and Porges, N.: *Venoms,* AAAS Publication 44. American Association for Advancement of Science, Washington, D. C., 1956, pp. 291-293.

Jensen, O. M.: Sudden death due to stings from bees and wasps. *Acta Pathol Micro Scand, 54:*9, 1962.

Jimenez-Porras, J. M.: Intraspecific variations in composition of venom of the jumping viper, *Bothrops nummifera. Toxicon, 2:*187, 1964.

Jimenez-Porras, J. M.: Pharmacology of peptides and proteins in snake venoms. *Ann Rev Pharmacol, 8:*299, 1968.

Jimenez-Porras, J. M.: Biochemistry of snake venoms. *Clin. Tox, 3:*389, 1970.

Jones, David L., and Miller, Joseph H.: Pathology of the dermatitis produced by the urticating caterpillar, *Automeris io. Arch Dermatol, 79:*81, 1959.

Jones, E. C.: *Tremoctopus violaceus* uses *Physalia* tentacles as weapons. *Science, 139:*764, 1963.

Jouannet, M.: L'Analyse immuno-electrophoretique appliquee aux venins de serpents. *Toxicon, 5:*191, 1968.

Juratsch, C. E., and Russell, F. E.: Immunological studies on snakes injected with *Crotalus* venom. *Herpeton, 6:*1, 1971.

Jutzy, D. A.; Biber, S. H.; Elton, N. W., and Lowry, E. C.: A clinical and pathological analysis of snake bites on the Panama Canal Zone. *Am J Trop Med Hyg, 2:*129, 1953.

Kaire, G. H.: Isolation of tick paralysis toxin from *Ixodes holocyclus. Toxicon, 4:*91, 1966.

Kaiser, Erich: Enzymatic activity of spider venoms. In Buckley, E. E., and Porges, N. (Ed.): *Venoms,* AAAS Publication 44. American Association for Advancement of Science, Washington, D. C., 1956, pp. 91-93.

Kaiser, E., and Rabb, W.: Collagenolytic activity of snake and spider venoms. *Toxicon, 4:*251, 1967.

Kanwar, K. C., and Sethi, R. C.: Sudanophilic and PAS-positive granules in the venom gland of *Vespa orientalis. Toxicon, 9:*179, 1971.

Kao, C. Y.: Tetrodotoxin, saxitoxin, and their significance in the study of excitation phenomena. *Pharmacol Rev, 18:*997, 1966.

Karlsson, E.; Eaker, D.L., and Porath, J.: Purification of a neurotoxin from the venom of *Naja nigricollis. Biochim Biophys Acta, 127:*505, 1966.

Karlsson, E., and Eaker, D.: Isolation of the principal neurotoxins of *Naja naja* subspecies from the Asian mainland. *Toxicon, 10:*217, 1972.

Karlsson, E.; Eaker, D., and Ryden, L.: Purification of a presynaptic neurotoxin from the venom of the Australian tiger snake *Notechis scutatus scutatus. Toxicon, 10:*405, 1972.

Karunaratne, K., and Panabokke, R. G.: Sea snake poisoning—case report. *J Trop Med Hyg, 75:*91, 1972.

Keegan, Hugh L., and Lockwood, W. R.: Secretory epithelium in venom glands of two species of scorpion of the genus *Centruroides. Am J Trop Med Hyg, 20:*770, 1971.

Keen, T. E. B.: Comparison of tentacle extracts from *Chiropsalmus quadrigatus* and *Chironex fleckeri. Toxicon, 9:*249, 1971.

Keen, T. E. B., and Crone, H. D.: Dermatonecrotic properties of extracts from the tentacles of the cnidarian *Chironex fleckeri. Toxicon, 7:*173, 1969.

Kellaway, C. H.: The immunity of Australian snakes to their own venoms. *Med J Aust, 2:*35, 1931.

Kellaway, C. H.: Symptomatology and treatment of the bites of Australian snakes. *Med J Aust, 2:*585, 1938.

Kellaway, C. H., and LeMessurier, D. H.: The venom of the platypus *(Ornithorhynchus anatimus). Aust J Exp Biol Med Sci, 13:*205, 1935.

Klauber, Laurence M.: *Rattlesnakes: their habits, life histories, and influence on mankind.* Berkeley and Los Angeles, University of California Press, vols. I, II, 1956.

Klein, W. E., and Bradshaw, R. H.: Portuguese man-of-war sting. *U. S. Armed Forces Med J, 2*:509, 1951.

Kniker, W. T.; Morgan, P. N.; Flanigan, W. J.; Reagan, P. W., and Dillaha, C. J.: An inhibitor of complement in the venom of the brown recluse spider. *Proc. Soc Exptl Biol Med, 131*:1432, 1969.

Knox, R.: Observations on the anatomy of the duckbilled animal of N. S. W., the *Ornithorhynchus paradoxus* of naturalists. *Mem Wernerian Soc Nat Hist, 5*:26, 1824.

Kocholaty, W. F.; Ledford, E. B.; Daly, J. G., and Billings, T. A.: Toxicity and some enzymatic properties and activities in the venoms of *Crotalidae, Elapidae,* and *Viperidae. Toxicon, 9*:131, 1971.

Kochva, Elazar: A quantitative study of venom secretion by *Vipera palaestinae. Am J Trop Med Hyg, 9*:381, 1960.

Kochva, Elazar, and Gans, Carl: The venom gland of *Vipera palaestinae* with comments on the glands of some other viperines. *Acta Anat, 62*:365-401, 1965.

Kochva, E., and Gans, C.: Histology and histochemistry of the venom gland of some crotaline snakes. *Copeia* (3) , 506, 1965.

Kochva, E.; Shayer-Wollberg, M., and Sobol, R.: The special pattern of the venom gland in *Atractaspis* and its bearing on the taxonomic status of the genus. *Copeia* (4): 763, 1967.

Kochva, E., and Gans, C.: Salivary glands of snakes. *Clin Tox, 3*:363, 1970.

Kohn, A. J.: Venomous marine snails of the genus *Conus.* In Keegan, H. L., and Macfarlane, W. V.: *Venomous and Poisonous Animals and Noxious Plants of the Pacific Region.* Oxford, Pergamon Press, 1963, pp. 83-96.

Kondo, S.; Sadahiro, S.; Yamauchi, K.; Kondo, H., and Murata, R.: Preparation and standardization of toxoid from the venom of *Trimeresurus flavoviridis* (Habu) . *Jap J Med Sci Biol, 24*:281, 1971.

Kondo, H.; Kondo, S., and Sadahiro, S.: Estimation by a new method of the amount of venom ejected by a single bite of *Trimeresurus* species. *Jap J Med Sci Biol, 25*:123, 1972.

Kurashiga, S.; Hara, Y.; Kawakami, M., and Mitsuhashi, S.: Studies on habu snake venom. VII. Heat stable myolytic factor and development of its activity by phospholipase A. *Jap J Microbiol, 10*:23, 1966.

Lakier, J. B., and Fritz, V. U.: Consumptive coagulopathy caused by a boomslang bite. *S Afr Med J, 43*:1052, 1969.

Lane, C. E.: The toxin of Physalia nematocysts. *Ann N Y Acad Sci, 90*:742, 1960.

Lane, Charles E.: Recent observations on the pharmacology of *Physalia* toxin. In Russell, F. E., and Saunders, P. R. (Eds.): *Animal Toxins.* Oxford, Pergamon Press, 1967, pp. 131-136.

Langlois, C.; Shulman, S., and Arbesman, C. E.: The allergic response to stinging insects. II. Immunologic study of human sera from allergic individuals. *J Allergy, 36*:12, 1965.

Latifi, M., and Shamloo, K. D.: Characteristic electrophoretic patterns of serum proteins of several species of snakes of Iran. *Canad J Biochem, 43:* 459, 1965.

Lebez, D.; Maretic, Z., and Kristan, J.: Studies on labeled animal poisons. I—Distribution of p^{32} labeled *Latrodectus tredecimguttatus* venom in the guinea pig. *Toxicon, 2:*251, 1965.

Ledbetter, E. O., and Kutscher, A. E.: The aerobic and anaerobic flora of rattlesnake fangs and venom. *Arch Environ Health, 19:770, 1969.*

Lee, C. Y.: Elapid neurotoxins and their mode of action. *Clin Toxicol, 3:* 457, 1970.

Lee, C. Y.; Chang, C. C.; Chiu, T. H.; Chiu, P. J.; Tseng, T. C., and Lee, S. Y.: Pharmacological properties of cardiotoxin isolated from Formosan cobra venom. *Nauyn-Schmiedberg Arch Pharmak Exp Pathol, 259:*360, 1968.

Leffkowitz, M.: On bites and stings in Israel. *Folia Medicina, 21:*5, 1962.

Lehmann, C. F.; Pipkin, J. L., and Ressmann, A. C.: Blister beetle dermatosis. *Arch Dermatol Syph, 71:*36, 1955.

Leopold, R. S., and Merriam, T. W.: The effectiveness of tourniquet, incision and suction in snake venom removal. *U S Nav Med Field Res Lab Res Proj MR 005. 09-0020, 1.3:211,* 1960.

Licht, L. E.: Death following possible ingestion of toad eggs. *Toxicon, 5:* 141, 1967.

Lieske, H.: Symptomatik und Therapie von Giftschlangenbissen. In Behringwerke Mitteilungen *Die Giftschlangen der Erde.* Marburg-Lahn, N. G. Elwert, 1963, pp. 121-160.

Lim, B. L., and Davie, C. E.: The bite of a bird-eating spider *Lampropelma violaceopedes. Med J Malaya, 24:*311, 1970.

Lockhart, W. E.: Pitfalls in rattlesnake bite. *Tex Med, 66:*42, 1970.

Loeb, L.; Alsberg, C. L.; Cooke, E.; Corson-White, E.; Fleisher, M. S.; Fox, H.; Githens, T. S.; Leopold, S.; Myers, M. K.; Rehfuss, M. E.; Rivas, D., and Tuttle, L.: *The Venom of Heloderma. Carnegie Inst Wash Publ. 177:* 1-244, 1913.

Longnecker, H. E.; Hurlbut, W. P.; Mauro, Alex, and Clark, A. W.: Effects of black widow spider venom on the frog neuromuscular junction. *Nature, 225:*701, 1970.

Lounsberry, C. R.: Rattlesnake anaphylaxis associated with a generalized dermatitis. *Arch Dermatol Syph, 29:*658, 1934.

Loveless, Mary H.: Immunization in wasp-sting allergy through venom-repositories and periodic insect stings. *J Immunol, 89:*204, 1962.

Loveless, M. H.: Immunization with native incriminated wasp venom in sting allergy: 12-year study of safety and effectiveness of a single visit yearly course. *Fed Proc, 27:*368, 1968.

Lubke, K.; Matthes, S. and Kloss, G.: Isolation and structure of N^a Formyl melittin. *Experientia, 27:*765, 1971.

Lutz, B.: Venomous toads and frogs. In Bücherl, W.; Buckley, E. E., and Deulofeu, V.: *Venomous Animals and their Venoms.* New York, Academic Press, 1971, vol. II, pp. 423-473.

Lyons, W. J.: Profound thrombocytopenia associated with *Crotalus ruber ruber* envenomation: a clinical case. *Toxicon, 9:*237, 1971.

McCollough, N. C., and Gennaro, J. F.: Evaluation of venomous snake bite in the southern United States from parallel clinical and laboratory investigations. *J Fla Med Assoc, 49:*959, 1963.

McCollough, N. C., and Gennaro, J. F.: Treatment of venomous snakebite in the United States. *Clin Toxicol, 3:*483, 1970.

MacConnell, J. G.; Blum, M. S., and Fales, H. M.: The chemistry of fire ant venom. *Tetrahedron, 26:*1129, 1971.

McCormick, W. F.: Fatal anaphylactic reactions to wasp sting. *Amer J Clin Pathol, 39:*485, 1963.

McCrone, J. D.: Comparative lethality of several *Latrodectus* venoms. *Toxicon, 2:*201, 1964.

McCrone, J. D., and Levi, H. W.: North American widow spiders of the *Latrodectus curacaviensis* group. *Psyche, 71:*12, 1964.

McCrone, J. D., and Netzloff, M. L.: An immunological and electrophoretical comparison of the venoms of the North American *Latrodectus* spiders. *Toxicon, 3:*107, 1965.

McCrone, John D., and Porter, R. J.: Hypertensive effect of *Latrodectus* venoms. *Q J Fla Acad Sci, 27:*307, 1965.

McCrone, John D., and Hatala, R. J.: Isolation and characterization of a lethal component from the venom of *Latrodectus mactans mactans.* In Russell, F. E., and Saunders, P. R.: *Animal Toxins,* Pergamon Press, Oxford, 1967, pp. 29-34.

McDonald, N. M., and Cottrell, G. A.: Purification and mode of action of toxin from *Eledone cirrhosa. Comp Gen Physiol, 3:*243, 1972.

McGovern, J. P.; Barkin, G. D.; McElhenny, T. R., and Wende, R.: *Megalopyge opercularis,* observations on its life history, natural history of its sting in man, and report of an epidemic. *J A M A, 175:*737, 1961.

Mackay, N.; Ferguson, J. C.; Bagshawe, A.; Forrester, A. T. T., and McNicol, G. P.: The venom of the boomslang *(Dispholidus typus) in vivo* and *in vitro* studies. *Thrombosis, 21:*234, 1969.

Mackinnon, J. E., and Witkind, J.: Aracnidismo necrotico. *Ann Fac Med Montevideo, 38:*76, 1953.

Macchaiavello, A.: La *Loxosceles laeta* cause del aracnidismo cutaneo o mancha gangrenosa de Chile. *Rev. Chile Hist Nat, 41:*11, 1937.

Maeno, H.; Mitsuhashi, S., and Sato, R.: Studies on habu snake venom. 2c. Studies of HB proteinase of habu venom. *Jap J Microbiol, 4:*173, 1960.

Maeno, H. S.; Mitsuhashi, T.; Okonogi, T.; Hoshi, S., and Homma, M.: Studies in snake venom. V. Myolysis caused by phospholipase A. *Jap J Microbiol, 10:*23, 1962.

Maguire, E. J.: *Chironex fleckeri* (sea wasp) sting. *Med J Aust, 2*:1137, 1968.

Malz, S.: Snake-bite in pregnancy. *J Obstet Gynaecol Br Comm, 74*:935, 1967.

Mann. G. T., and Bates, H. R.: The pathology of insect bites. *South Med J, 53*:1399, 1960.

Maretic, Z. Erfahrungen mit Stichen von Giftfishchen; *Acta Trop, 14*:157, 1957.

Maretic, Z.: *Chiracanthium punctorium* Villiers—eine europaische Giftspinne. *Med Klin, 57:* 1576, 1962.

Maretic, Z.: Electrocardiographic changes in man and experimental animals provoked by the venom of *Latrodectus tredecimguttatus*. *Toxicon, 1*:127, 1963.

Maretic, Z.: Latrodectism. *Jugoslav Akad Znanosti Umjetnosti, 17*:63, 1966.

Maretic, Z.: Venom of an east African orthognath spider. In Russell, F. E., and Saunders, P. R. (Eds.): *Animal Toxins*. Oxford, Pergamon Press, 1967, pp. 23-28.

Maretic, Z., and Lebez, D.: *Lycosa tarentula* in fact and fiction. *Bull Mus Nation Hist Nat, 41*:260, 1970.

Maretic, Z.; Levi, H. W., and Levi, L. R.: The theridiid spider, *Steatoda paykulliana*, poisonous to mammals. *Toxicon, 2*:149, 1964.

Marr, J. J.: Portuguese man-of-war envenomization. A personal experience. *J A M A, 199*:115, 1967.

Marsh, Helene: Preliminary studies of the venoms of some vermivorous Conidae. *Toxicon, 8*:271, 1970.

Marsh, Helene: The caseinase activity of some vermivorous cone shell venoms. *Toxicon, 9*:63, 1971.

Martin, C. J., and Tidswell, F.: Observations on the femoral gland of Ornithorhynchus and its secretion together with an experimental enquiry concerning its supposed toxic action. *Proc Linnean Soc N S Wales, 9*:471, 1895.

Master, R. W.; Rao, S., and Soman, P. D.: The electrophoretic separation of scorpion venoms. *Biochem Biophys Acta, 71*:422, 1963.

Maynard, C. J.: Singular effects produced by bite of shorttailed shrew. *Contrib Sci, 1*:57, 1889.

Mazzotti, Luis, and Bravo-Becherelle, M. A.: Scorpionism in the Mexican republic. In Keegan, H. L. and Macfarlane, W. V.: *Venomous and Poisonous Animals and Noxious Plants of the Pacific Region*. Oxford, Pergamon Press, 1963, pp. 119-131.

Meaume, J.: Les venins des serpents agents modificateurs de la coagulation sanguine. *Toxicon, 4*:25, 1966.

Mebs, D.: Proteolytische Aktivat eines Vogelspinnerengiftes. *Naturwissenschaft, 57*:308, 1970.

Mebs, D.: Biochemistry of *Heloderma* venom. In DeVries, A., and Kochva, E.: *Toxins of Animal and Plant Origin*. New York, Gordon and Breach, 1972, vol. II, pp. 499-513.

speces de scorpions Nord-Africans. *Toxicon, 2:*51, 113, 123, 1964.

anda, F.; Rochat, I.; Rochat, C., and Lissitzky: Essais de purification les neurotoxines du venin d'un scorpion d'Amerique du Sud *(Tityus serulatus)* par des methodes chromatographiques. *Toxicon, 4:*145, 1966.

chell, J. H.: Eye injuries due to jelly fish. *Med J Aust, 2:*303, 1962.

rakul, C. and Impun, C.: The hemorrhagic phenomena associated with green pit viper *(Tremeresurus erythurus* and *T. popeorum* bites in children. *Clin Pediat 12 —*215, 1973.

hamed, A. H.; El-Serougi, M., and Hanna, M. M.: Observations on the effects of *Echis carinatus* venom on blood clotting. *Toxicon, 6:*215, 1969a.

hamed, A. H.; Kamel, A., and Ayobe, M. H.: Studies of phospholipase A and B activities of Egyptian snake venoms and a scorpion toxin. *Toxicon, 6:*293, 1969b.

hamed, A. H.; El-Serougi, M., and Khaled, L. Z.: Effect of *Cerastes cerastes* venom on blood coagulation mechanisms. *Toxicon, 7:*181, 1969c.

hamed, A. H.; El-Se rougi, M. S., and Hamed, R. M.: The effect of *Naja nigricollis* venom on blood clotting. *Toxicon, 9:*173, 1971.

hamed, A. H.; Darwish, M. A., and Hani-Ayobe, M.: Immunological studies on Egyptian cobra antivenin. *Toxicon, 11:*31, 1973a. Immunological studies on *Naja nigricollis* antivenin; 1973b. *Ibid.,* 35.

rhouse, C. H.: Unusual reaction to ant bites. *J A M A, 141:*193, 1949.

roz, C.; deVries, A. and Sela, M.: Isolation and characterization of a neurotoxin from *Vipera palaestinae* venom. *Biochem Biophys Acta, 124:* 136, 1966.

seley, T.: Coral snake bite: recovery following symptoms of respiratory paralysis. *Ann Surg, 163:*943, 1966.

sher, H. S.; Fuhrman, F. A.; Buchwald, H. D., and Fischer, H. G.: Tarchatoxin-Tetrodotoxin: a potent neurotoxin. *Science, 144:*1100, 1964.

eller, H. L.: Further experiences with severe allergic reactions to insect stings. *New Eng J M ed, 261:*374, 1959.

llanney, P. J.: Treatment of sting ray wounds. *Clin Toxicol, 3:*613, 1970.

iller-Eberhard, H. J. and Fjellström, K.: Isolation of the anticomplementary protein from cobra venom and its mode of action on C3. *J Immunol, 107:*1666, 1971.

ndle, P. M.: Scorpion stings. *Br Med J, 1:*1042, 1961.

njal, D., and Elliot, W. B.: Studies of antigenic fractions in honey-bee *(Apis mellifera)* venom. *Toxicon, 9:*229, 1971.

njal, D., and Elliot, W. B.: Immunological and histochemical identity of esterases and other antigens in elapid venoms. *Toxicon, 10:*47, 1972.

rnaghan, M. F.: Site and mechanism of tick paralysis. *Science, 131:*418, 1960.

kai, K.; Nakai, C., and Hayashi, K.: Purification and some properties of toxin A from the venom of the Indian cobra *(Naja naja). Jap J Med*

References

Mebs, D., and Raudonat, H. W.: Biochemical investigatior
   venom. *Mem Inst Butantan Simp Internac, 33*:907, 1966
Meldrum, B. S.: The actions of snake venom on nerve ;
   pharmacology of phospholipase A and polypeptide t
   Rev, 17:393, 1965.
Mendelssohn, H.; Golani, I., and Marder, U.: Agricultural
   the distribution of venomous snakes and snake bite in I
   A., and Kochva, E.: *Toxins of Animal and Plant O.*
   Gordon and Breach, 1971, vol. I, pp. 2-15.
Mendes, E.; Cintra, A. U., and Correa, A.: Allergy to sna
   lergy, 31:68, 1960.
Mendes, F. G.; Abbud, L., and Umiji, S.: Cholinergic ac
   ates of sea urchin pedicellariae. *Science, 139*:408, 1963
Meyer, K., and Linde, H.: Collection of toad venoms and
   toad venom steroids. In Bücherl, W.; Buckley, E. E., ;
   Venomous Animals and their Venoms. New York, Aca(
   vol. II, pp. 521-556.
Micheel, F.; Dietrich, H., and Bischoff, G.: Uber die Neurc
   von Cobraarten. *Zeit Physiol Chem, 249*:157, 1937.
Micks, D. W.: Clinical effects of the sting of the "puss c
   lopyge opercularis) on man. *Tex Rpt Biol Med, 10*:39§
Micks, D. W.: Insects and other arthropods of medical imp
   Tex Rpt Biol Med, 18:624, 1960.
Middlebrook, R. E.; Wittle, L. W.; Scura, E. D., and Lan
   and purification of a toxin from *Millepora dichotoma*
   1971.
Minton, S. A.: Snakebite in the midwestern region. *Q I*
   Med Ctr, 14:28, 1952.
Minton, S. A.: Observations on toxicity and antigenic n
   from juvenile snakes. In Russell, F. E., and Saunders,
   mal Toxins, Oxford, Pergamon Press, 1967, pp. 211-
Minton, S. A.: Preliminary observations on the venom of
   (Trimeresurus wagleri). Toxicon, 6:93, 1968a.
Minton, S. A.: Antigenic relationships of the venom of
   lepidota to that of other snakes. *Toxicon, 6*:59,
Minton, S. A.: Identification of poisonous snakes. *Clin 1*
Minton, S. A.: Poisonous spiders of Indiana with a re
   Chiracanthium mildei. J Indiana State Med Assoc,
Minton, S. A.: Common antigens in snake sera and veno
   and Kochva, E. (eds.) *Toxins of Animal and Plant*
   Gordon and Breach vol. III, pp. 903-917, 1973.
Minton, S. A., and Minton, M. R.: *Venomous Reptiles.*
   ners, 1969.
Miranda, F.; Rochat, H., and Lissitzky, S.: Sur les net

Mebs, D., and Raudonat, H. W.: Biochemical investigations on *Heloderma* venom. *Mem Inst Butantan Simp Internac, 33:*907, 1966.

Meldrum, B. S.: The actions of snake venom on nerve and muscle. The pharmacology of phospholipase A and polypeptide toxins. *Pharmacol Rev, 17:*393, 1965.

Mendelssohn, H.; Golani, I., and Marder, U.: Agricultural development and the distribution of venomous snakes and snake bite in Israel. In DeVries, A., and Kochva, E.: *Toxins of Animal and Plant Origin.* New York, Gordon and Breach, 1971, vol. I, pp. 2-15.

Mendes, E.; Cintra, A. U., and Correa, A.: Allergy to snake venoms. *J Allergy, 31:*68, 1960.

Mendes, F. G.; Abbud, L., and Umiji, S.: Cholinergic action of homogenates of sea urchin pedicellariae. *Science, 139:*408, 1963.

Meyer, K., and Linde, H.: Collection of toad venoms and chemistry of the toad venom steroids. In Bücherl, W.; Buckley, E. E., and Deulofeu, V.: *Venomous Animals and their Venoms.* New York, Academic Press, 1971, vol. II, pp. 521-556.

Micheel, F.; Dietrich, H., and Bischoff, G.: Uber die Neurotoxine aus Giften von Cobraarten. *Zeit Physiol Chem, 249:*157, 1937.

Micks, D. W.: Clinical effects of the sting of the "puss caterpillar" *(Megalopyge opercularis)* on man. *Tex Rpt Biol Med, 10:*399, 1952.

Micks, D. W.: Insects and other arthropods of medical importance in Texas. *Tex Rpt Biol Med, 18:*624, 1960.

Middlebrook, R. E.; Wittle, L. W.; Scura, E. D., and Lane, C. E.: Isolation and purification of a toxin from *Millepora dichotoma. Toxicon, 9:*333, 1971.

Minton, S. A.: Snakebite in the midwestern region. *Q Bull Indiana Univ Med Ctr, 14:*28, 1952.

Minton, S. A.: Observations on toxicity and antigenic makeup of venoms from juvenile snakes. In Russell, F. E., and Saunders, P. R. (Eds.): *Animal Toxins,* Oxford, Pergamon Press, 1967, pp. 211-222.

Minton, S. A.: Preliminary observations on the venom of Wagler's pit viper *(Trimeresurus wagleri). Toxicon, 6:*93, 1968a.

Minton, S. A.: Antigenic relationships of the venom of *Atractaspis microlepidota* to that of other snakes. *Toxicon, 6:*59, 1968b.

Minton, S. A.: Identification of poisonous snakes. *Clin Tox, 3:*347, 1970.

Minton, S. A.: Poisonous spiders of Indiana with a report of a bite by *Chiracanthium mildei. J Indiana State Med Assoc, 65:*426, 1972.

Minton, S. A.: Common antigens in snake sera and venoms. In DeVries, A. and Kochva, E. (eds.) *Toxins of Animal and Plant Origin.* New York Gordon and Breach vol. III, pp. 903-917, 1973.

Minton, S. A., and Minton, M. R.: *Venomous Reptiles.* New York. Scribners, 1969.

Miranda, F.; Rochat, H., and Lissitzky, S.: Sur les neurotoxines de deux especes de scorpions Nord-Africans. *Toxicon, 2:*51, 113, 123, 1964.

Miranda, F.; Rochat, H.; Rochat, C., and Lissitzky: Essais de purification des neurotoxines du venin d'un scorpion d'Amerique du Sud *(Tityus serrulatus)* par des methodes chromatographiques. *Toxicon, 4:*145, 1966.

Mitchell, J. H.: Eye injuries due to jelly fish. *Med J Aust, 2:*303, 1962.

Mitrakul, C. and Impun, C.: The hemorrhagic phenomena associated with green pit viper *(Trimeresurus erythurus* and *T. popeorum* bites in children. *Clin Pediat 12:*215, 1973.

Mohamed, A. H.; El-Serougi, M., and Hanna, M. M.: Observations on the effects of *Echis carinatus* venom on blood clotting. *Toxicon, 6:*215, 1969a.

Mohamed, A. H.; Kamel, A., and Ayobe, M. H.: Studies of phospholipase A and B activities of Egyptian snake venoms and a scorpion toxin. *Toxicon, 6:*293, 1969b.

Mohamed, A. H.; El-Serougi, M., and Khaled, L. Z.: Effect of *Cerastes cerastes* venom on blood coagulation mechanisms. *Toxicon, 7:*181, 1969c.

Mohamed, A. H.; El-Serougi, M. S., and Hamed, R. M.: The effect of *Naja nigricollis* venom on blood clotting. *Toxicon, 9:*173, 1971.

Mohamed, A. H.; Darwish, M. A., and Hani-Ayobe, M.: Immunological studies on Egyptian cobra antivenin. *Toxicon, 11:*31, 1973a. Immunological studies on *Naja nigricollis* antivenin; 1973b. *Ibid.*, 35.

Morhouse, C. H.: Unusual reaction to ant bites. *J A M A, 141:*193, 1949.

Moroz, C.; deVries, A., and Sela, M.: Isolation and characterization of a neurotoxin from *Vipera palaestinae* venom. *Biochem Biophys Acta, 124:* 136, 1966.

Moseley, T.: Coral snake bite: recovery following symptoms of respiratory paralysis. *Ann Surg, 163:*943, 1966.

Mosher, H. S.; Fuhrman, F. A.; Buchwald, H. D., and Fischer, H. G.: Tarchatoxin-Tetrodotoxin: a potent neurotoxin. *Science, 144:*1100, 1964.

Mueller, H. L.: Further experiences with severe allergic reactions to insect stings. *New Eng J Med, 261:*374, 1959.

Mullanney, P. J.: Treatment of sting ray wounds. *Clin Toxicol, 3:*613, 1970.

Müller-Eberhard, H. J. and Fjellström, K.: Isolation of the anticomplementary protein from cobra venom and its mode of action on C3. *J Immunol, 107:*1666, 1971.

Mundle, P. M.: Scorpion stings. *Br Med J, 1:*1042, 1961.

Munjal, D., and Elliott, W. B.: Studies of antigenic fractions in honey-bee *(Apis mellifera)* venom. *Toxicon, 9:*229, 1971.

Munjal, D., and Elliott, W. B.: Immunological and histochemical identity of esterases and other antigens in elapid venoms. *Toxicon, 10:*47, 1972.

Murnaghan, M. F.: Site and mechanism of tick paralysis. *Science, 131:*418, 1960.

Nakai, K.; Nakai, C., and Hayashi, K.: Purification and some properties of toxin A from the venom of the Indian cobra *(Naja naja). Jap J Med*

Sci Biol, 24:27, 1971.

Neumann, W., and Habermann, E.: Paper electrophoresis separation of pharmacologically and biochemically active components of bee and snake venom. In Buckley, E. E., and Porges, N. (Eds.): *Venoms, AAAS* Publication 44. American Association for Advancement of Science, Washington, D. C., 1956, pp. 171-174.

Norment, B. R., and Vinson, S. B.: Effect of *Loxosceles reclusa* venom on *Heliothis virescens* larvae. *Toxicon, 7*:99, 1969.

O'Brien, J. R.: The effect of cobra venom and bee venom on plasma. Some evidence on the possible chemical composition of the Christmas factor. *Br J Haematol, 2*:430, 1956.

O'Conner, R.; Rosenbrook, W. Jr., and Erickson, R.: Hymenoptera: pure venom from bees, wasps, and hornets. *Science, 139*:420, 1963.

O'Conner, R.; Rosenbrook, W. Jr., and Erickson, R.: Disc electrophoresis of hymenoptera venoms and body proteins. *Science, 145*:1320, 1964a.

O'Conner, R.; Sher, R. A.; Rosenbrook, W. Jr., and Erickson, B. A.: Death from "wasp" sting. *Ann Allergy, 22*:385, 1964b.

O'Conner, R., and Erickson, R.: Hymenoptera antigens: an immunological comparison of venoms, venom sac extracts, and whole-insect extracts. *Ann Allergy, 23*:151, 1965.

O'Conner, R.; Henderson, G.; Nelson, D.; Parker, R., and Peck, M. L.: The venom of the Honeybee *(Apis mellifera)* I. general character. In Russell, F. E., and Saunders, P. R. (Eds.): *Animal Toxins.* Oxford, Pergamon Press, 1967, pp. 17-22.

Ohsaka, A.; Omori-Satoh, T.; Kondo, H.; Kondo, S., and Murata, R.: Biochemical and pathological aspects of hemorrhagic principles in snake venoms with special reference to habu *(Trimeresurus flavoviridis)* venom. *Mem Inst Butantan Simp Internac, 33*:193, 1966.

Okabe, K.; Sugimura, T., and Kasuga, T.: Inactivation of Friend virus by snake venom. *Gann, 55*:19, 1964.

Omari-Satoh, T., and Ohsaka, A.: Purification and some properties of hemorrhagic principle 1 in the venom of *Trimeresurus flavoviridis. Biochim Biophys Acta, 207*:432, 1970.

Oshima, G.; Sato-Ohmori, T., and Suzuki, T.: Proteinase, arginineester hydrolase and a kinin releasing enzyme in snake venoms. *Toxicon, 7*:229, 1969.

Ouyang, C., and Teng, C.: Purification and properties of the anticoagulant principle of *Agkistrodon acutus* venom. *Biochim Biophys Acta, 278*:155, 1972.

Pacy, H.: Australian catfish injuries with report of a typical case. *Med J Aust, 2*:63, 1966.

Padgett, F., and Levine, A. S.: The fine structure of Rauscher leukemia virus as revealed by incubation in snake venom. *Virology, 30*:623, 1966.

Parnas, Itzchak, and Russell, F. E.: Effects of venoms on nerve, muscle and neuromuscular junction. In Russell, F. E., and Saunders, P. R. (Eds.): *Animal Toxins*. Oxford, Pergamon Press, 1967, pp. 401-415.

Parrish, H. M.: Mortality from snakebites. *Publ Health Rep, 72*:1027, 1957a.

Parrish, H. M.: Survey of human allergy to North American snake venoms. *Ariz Med, 14*:461, 1957b.

Parrish, H. M.: Analysis of 460 fatalities from venomous animals. *Am J Med Sci, 245*:129, 1963.

Parrish, H. M.: Incidence of treated snakebites in the United States. *Pub Hlth Rep, 81*:269, 1966.

Parrish, H. M.; MacLaurin, A. W., and Tuttle, R. L.: North American pit vipers: bacterial flora of the mouths and venom glands. *Virginia Med Month, 83*:383, 1956.

Parrish, H. M.; Goldner, J. C., and Silberg, S. L.: Poisonous snakebites causing no venenation. *Postgrad Med, 39*:265, 1966.

Parrish, H. M., and Khan, M. S.: Snakebite during pregnancy; *Obstet Gynecol, 27*:468, 1966.

Patterson, Robert A.: Action of scorpion venom. *Am J Trop Med Hyg, 9*:410, 1960.

Patterson, R. A.: Some physiological effects caused by venom from the Gila Monster, *Heloderma suspectum. Toxicon, 5*:5, 1967a. Smooth muscle stimulating action of venom from the Gila Monster, *Heloderma suspectum, Ibid.*, 11, 1967b.

Patterson, R. A., and Lee, I. S.: Effects of *Heloderma suspectum* venom on blood coagulation. *Toxicon, 7*:321, 1969.

Pearn, J. H.: Survival after snake-bite with prolonged neurotoxic envenomation. *Med J Aust, 2*:259, 1971.

Pearson, O. P.: The submaxillary glands of shrews. *Anat Rec, 107*:61, 1950.

Pearson, O. P.: A toxic substance from the salivary glands of a mammal (short-tailed shrew). In Buckley, E. E., and Porges, N. (Eds.): *Venoms,* AAAS Publication 44. American Association for Advancement of Science, Washington, D. C., 1956, pp. 55-58.

Pereire Lima, F. A.; Schenberg, S.; Shiripa, L. N., and Nagamori, A.: ATPase and phosphodiesterase differentiation in snake venoms. In DeVries, A., and Kochva, E.: *Toxins of Animal and Plant Origin.* New York, Gordon and Breach, 1971, vol. I, p. 463.

Perkash, A., and Sarup, B. M.: Red cell abnormalities after snake bite. *J Trop Med Hyg, 75*:85, 1972.

Perlman, F.: Near fatal reactions to bee and wasp stings: a review and report of seven cases. *J Mt Sinai Hosp, 22*:336, 1955.

Perlman, F.: Insects as allergen injectants. *Calif Med, 96*:1, 1962.

Pesce, H., and Delgado, A.: Lepidopterismo y erucismo. *Mem Inst Butantan Simp Internac, 33*:829, 1966.

Phelps, D. R.: Stone-fish poisoning. *Med J Aust, 1*:293, 1960.

Phisalix, M.: *Animaux Venimeux et Venins*. Masson, Paris, 1922.

Philpot, V. B. Jr.: Neutralization of snake venom *in vitro* by serum from the nonvenomous snake *Elaphe quadrivirgata*. *Herpetologica, 10:*158, 1954.

Philpot, V. B. Jr., and Smith, R. G.: Neutralization of pit viper venom by king snake serum. *Proc Soc Exptl Biol Med, 74:*521, 1950.

Picado, C.: Estudo experimental sobre o veneno de *Lethocerus del-ponti* (de Carlo). *Mem Inst Butantan, 10:*303, 1936.

Piek, T., and Thomas, R. T.: Paralysing venoms of solitary wasps. *Comp Biochem Physiol, 30:*13, 1969.

Pigulevsky, S. V., and Michaleff, P. V.: Poisoning by the medusa, *Gonionemus vertens*. *Toxicon, 7:*145, 1969.

Pisano, J. J.: Wasp kinin. *Mem Inst Butantan Simp Internac, 33:*441, 1966.

Poen-King, T.: Myocarditis from scorpion stings. *Br Med J, 1:*374, 1963.

Pucek, M.: Chemistry and pharmacology of insectivore venoms. In Bücherl, W.; Buckley, E. E., and Deulofeu, V.: *Venomous Animals and their Venoms*. New York, Academic Press, 1968, vol. I, pp. 43-50.

Puranananda, C.; Lauhatirananda, P., and Ganthavorn, S.: Cross immunological reactions in snake venoms. *Mem Inst Butantan Simp Internac, 33:* 327, 1966.

Rabb, G.: Toxic salivary glands in the primitive insectivore *Solenodon*. *Nat Hist Miscellanea*, No. 170, 1959.

Raitano, A. C.: Studies on the enzymatic effects of cobra venom on Rauscher murine leukemia virus. Unpublished Ph.D. dissertation, Indiana University, 1968.

Ramsey, H. W.; Snyder, G. K.; Kitchen, H., and Taylor, W. J.: Fractionation of coral snake venom. Preliminary studies on the separation and characterization of the protein fractions. *Toxicon, 10:*67, 1972.

Rathjin, W. F., and Halstead, B. W.: Report on two fatalities due to stingrays. *Toxicon, 6:*301, 1969.

Rathmayer, Werner: The effect of the poison of spider and digger-wasps on their prey. *Mem Inst Butantan Simp Internac, 33:*651, 1966.

Reid, H. Alistair: Sea-snake bite research. *Trans Roy Soc Trop Med Hyg, 50:*517, 1956a.

Reid, H. A.: Sea snake bites. *Br Med J, 1:*73, 1956b.

Reid, H. A.: Sea snake bite: a survey of fishing villages in northwest Malaya. *Br Med J, 1:*1266, 1957a.

Reid, H. A.: Antivenene reaction following accidental sea snake bite. *Br Med J, 2:*26, 1957b.

Reid, H. A.: Diagnosis, prognosis, and treatment of sea-snake bite. *Lancet, 2:*399, 1961.

Reid, H. A.: Epidemiology of snake bite in north Malaya. *Br Med J, 1:*992, 1963.

Reid, H. A.: Cobra bites. *Br Med J, 2:*540, 1964.

Reid, H. A.: Defibrination by *Agkistrodon rhodostoma* venom. In Russell,

F. E., and Saunders, P. R. (Eds.): *Animal Toxins*. Oxford Pergamon Press, 1967, p. 323-336.

Reid, H. A.: Snakebite in the tropics. *Br Med J, 3:*359, 1968a.

Reid, H. A.: Symptomatology, pathology, and treatment of land snake bite in India and southeast Asia. In Bücherl, W.; Buckley, E. E., and Deulofeu, V.: *Venomous Animals and their Venoms*. New York, Academic Press, 1968b, vol. I, pp. 611-642.

Reid, H. A.; Thean, P. C.; Chan, K. E., and Baharom, A. R.: Clinical effects of bites by Malayan viper *(Ancistrodon rhodostoma)*. *Lancet, 1:* 617, 1963a.

Reid, H. A.; Chan, K. E., and Thean, P. C.: Prolonged coagulation defect (defibrination syndrome) in Malayan viper bite. *Lancet, 1:*621, 1963b.

Reid, H. A.; Thean, P. C., and Martin, W. J.: Specific antivenene and prednisone in viper-bite poisoning: controlled trial. *Br Med J, 2:*1378, 1963c.

Reid, H. A., and Chan, K. E.: The paradox in therapeutic defibrination. *Lancet, 1:*485, 1968.

Remington, Charles L.: The bite and habits of a giant centipede *(Scolopendra subspinipes)* in the Philippine Islands. *Am J Trop Med, 30:*453, 1950.

Rice, R. D., and B. W. Halstead: Report of a fatal cone shell sting by *Conus geographus* Linnaeus. *Toxicon, 5:*223, 1968.

Ridley, H.: Snake venom ophthalmia. *Br J Ophthal, 28:*568, 1944.

Robertson, S. S. D., and Delpierre, G. R.: Studies on South African snake venoms—IV. Some enzymatic activities in the venom of the boomslang *Dispholidus typus*. *Toxicon, 7:*189, 1969.

Rochat, H.; Rochatt, C.; Miranda, F.; Lissitzky, S., and Edman, P.: The amino acid sequence of neurotoxin I of *Androctonus australis* Hector. *Europ J Biochem, 17:*262, 1970.

Rodrigues, Roberto J.: Pharmacology of South American freshwater stingray venom. *Trans N Y Acad Sci, 34:*677, 1972.

Ronka, E. K., and Roe, W. F.: Cardiac wound caused by the spine of the stingray. *Mil Surg, 97:*135, 1945.

Rose, I.: A review of tick paralysis. *Canad Med Assoc J, 70:*175, 1954.

Rosenberg, Herbert I.: Histology, histochemistry and emptying mechanism of the venom glands of some elapid snakes. *J Morphol, 123:*133, 1967.

Rosenbrook, W. Jr., and O'Conner, R.: The venom of the mud-dauber wasp I. *Sceliphron caementarium:* General character. *Can J Biochem, 42:*1567, 1964a. The venom of the mud-dauber wasp II. *Sceliphron caementarium:* Protein content. *Ibid.,* p. 1005, 1964b.

Rosenfeld, G.: Symptomatology, pathology and treatment of snake bites in South America. In Bücherl, W.; Buckley, E. E., and Deulofeu, V.: *Venomous Animals and their Venoms,* New York, Academic Press, 1971, vol. II, pp. 345-384.

Rosin, R.: Effects of the venom of the scorpion *Nebo hierichonticus* on white mice, other scorpions and paramecia. *Toxicon, 7:*71, 1969.

Rothschild, M.; Reichstein, T.; Von Euw, J.; Aplin, R., and Harman, R. R. M.: Toxic lepidoptera. *Toxicon, 8:*293, 1970.

Rowlands, J. B.; Mastaglia, F. L.; Kakulas, B. A., and Hainsworth, D.: Clinical and pathological aspects of a fatal case of mulga *(Pseudechis australis)* snakebite. *Med J Aust, 1:*226, 1969.

Russell, F. E.: Stingray injuries: a review and discussion of their treatment. *Am J Med Sci, 226:*611, 1953.

Russell, F. E.: Stingray injuries. *Pub Health Rep, 74:*855, 1959.

Russell, F. E.: Muscle relaxants in black widow spider *(Latrodectus mactans)* poisoning. *Am J Med Sci, 243:*159, 1962.

Russell, F. E.: Marine toxins and venomous and poisonous marine animals. *Adv Marine Biol, 3:*255, 1965.

Russell, F. E.: Physalia stings: a report of two cases. *Toxicon, 4:*65, 1966a.

Russell, F. E.: Phosphodiesterase of some snake and arthropod venoms. *Toxicon, 4:*153, 1966b.

Russell, F. E.: Metronidazol in snake venom poisoning; *Mem Inst Butantan Simp Internac, 33:*845, 1966c.

Russell, F. E.: Comparative pharmacology of some animal toxins. *Fed Proc, 26:*1206, 1967.

Russell, F. E.: Bite by the spider *Phidippus formosus:* case history. *Toxicon, 8:*193, 1970.

Russell, F. E., and Emery, J. A.: Incision and suction following injection of rattlesnake venom. *Am J Med Sci, 241:*160, 1961.

Russell, F. E., and van Harreveld, A.: Cardiovascular effects of the venom of the round stingray, *Urobatis halleri. Arch Internat Physiol, 62:*322, 1954.

Russell, F. E.; Barritt, W. C., and Fairchild, M. D.: Electrocardiographic patterns evoked by venom of the stingray. *Proc Soc Exptl Biol Med, 96:*634, 1957.

Russell, F. E.; Panos, T. C.; Kang, L. W.; Warner, A. M., and Colket, T. C.: Studies on the mechanism of death from stingray venom a report of two fatal cases. *Am J Med Sci, 235:*566, 1958a.

Russell, F. E.; Fairchild, M. D., and Michaelson, J.: Some properties of the venom of the stingray. *Med Arts Sci, 12:*78, 1958b.

Russell, F. E., and Emery, J. A.: Venom of the weevers *Trachinus draco* and *Trachinus vipera. Ann N Y Acad Sci, 90:*805, 1960.

Russell, F. E.; Buess, F. W.; Woo, M. Y., and Eventov, R.: Zootoxicological properties of venom L-amino acid oxidase. *Toxicon, 1:*229, 1963.

Russell, F. E., and Lauritzen, L.: Antivenins. *Trans Royal Soc Trop Med Hyg, 60:*797, 1966.

Russell, F. E.; Alender, C. B., and Buess, F. W.: Venom of the scorpion *Vejovis spinigerus. Science, 159:*90, 1968.

Russell, F. E.; Waldron, W. G., and Madon, M. B.: Bites by the brown spiders *Loxosceles unicolor* and *Loxosceles arizonica* in California and Arizona. *Toxicon, 7*:109, 1969.

Russell, F. E., and Buess, F. W.: Gel electrophoresis: a tool in systematics. Studies with *Latrodectus mactans* venom. *Toxicon, 8*:81, 1970.

Russell, F. E., and Puffer, H. W.: Pharmacology of snake venoms. *Clin Tox, 3*:432, 1970.

Samejima, Y.; Iwanaga, S.; Suzuki, T., and Kawauchi, S.: Studies on snake venom phospholipase A: isolation, characterization, and partial amino acid sequence. *Jap J Med Sci Biol, 24*:31, 1971.

Sanzenbacher, K. E., and Conrad, E.: Tick paralysis — a treatable killer neuropathy. *South Med J, 61*:764, 1968.

Sarkar, N. K.: Isolation of cardiotoxin from cobra venom *(Naja tripudians,* monocellate variety). *J Indian Chem Soc, 24*:227, 1947.

Sarkar, N. K., and Devi, A.: Enzymes in snake venoms. In Bücherl, W.; Buckley, E. E., and Deulofeu, V.: *Venomous Animals and their Venoms.* New York, Academic Press, 1968, vol. I, pp. 167-216.

Saunders, J., and Johnson, B. D.: *Hadrurus arizonensis* venom: a new source of acetylcholinesterase. *Am J Trop Med Hyg, 19*:345, 1970.

Saunders, Paul R.: Pharmacological and chemical studies of the venom of the Stonefish (Genus *Synanceja)* and other scorpion fishes. *Ann N Y Acad 90*:798, 1960.

Saunders, P. R., and Tokes, L.: Purification and properties of the lethal fraction of the venom of the stonefish *Synanceja horrida* (Linnaeus). *Biochem Biophys Acta, 52*:527, 1961.

Sawai, Y., and Kawamura, Y.: Study on the toxoids against the venoms of certain Asian snakes. *Toxicon, 7*:19, 1969.

Sawai, Y.; Kawamura, Y.; Fukuyama, T.; Okonogi, T., and Ebisawa, I.: Studies on the improvement of treatment of habu *(Trimeresurus flavoviridis)* bites VIII. A field trial of prophylactic immunization of habu venom toxoid. *Jap J Exp Med, 39*:197, 1969.

Sawai, Y.; Koba, K.; Okonogi, T.; Mishima, S.; Kawamura, Y.; Chinzei, H.; Ibrahim, A. B.; Devaraj, T.; Phong-Aksara, S.; Puranananda, C.; Salafranca, E. S.; Sumpaico, J. S.; Tseng, C. S.; Taylor, J. B.; Wu, C. S., and Kuo, T. P.: An epidemiological study of snakebites in southeast Asia. *Jap J Exp Med, 42*:283, 1971.

Sawaya, P.: Toxic marine invertebrates — venomous and noxious fishes of fresh water. *Mem Inst Butantan Simp Internac, 33*:31, 1966.

Schachter, M., and Thain, E. M.: Chemical and pharmacological properties of the potent, slow contracting substance (kinin) in wasp venom. *Br J Pharmacol, 9*:352, 1954.

Schaeffer, R. C.; Carlson, R. W., and Russell, F. E.: Some chemical properties of the venom of the scorpionfish *Scorpaena guttata*. *Toxicon, 9*:69, 1971.

Schaeffer, R. C. Jr.; Bernick, S.; Rosenquist, T. H., and Russell, F. E.: The histochemistry of the venom glands of the rattlesnake *Crotalus viridis helleri* I. Lipid and non-specific esterase. *Toxicon, 10:*183, 1972a. The histochemistry of the venom glands of the rattlesnake *Crotalus viridis helleri* II. Monamine oxidase and alkaline phosphatase. Ibid., 295, 1972b.

Schapel, G. J.; Utley, D., and Wilson, G. C.: Envenomation by the Australian common brown snake *Pseudonaja (Demansia) textilis. Med J Aust, 1:*142, 1971.

Schenberg, S.: Geographical pattern of crotamine distribution in the same rattlesnake species. *Science, 129:*1361, 1959.

Schenberg, S.: Immunological (Ouchterlony method) identification of intra-subspecies qualitative differences in snake venom composition. *Toxicon, 1:*67, 1963.

Schenberg, S., and Pereira Lima, F. A.: Pharmacology of the polypeptides from the venom of the spider *Phoneutria fera. Mem Inst Butantan Simp Internac, 33:*627, 1966.

Schenone, H., and Prats, F.: Arachnidism by *Loxosceles Laeta. Arch Dermatol, 83:*139, 1961.

Schieck, A.; Kornalik, F., and Habermann, E.: The prothrombin-activating principle from *Echis carinatus* venom. *Nauyn-Schmiedeberg's Arch Pharmak Exp Pathol, 272:*402, 1972.

Schöttler, W. H. A.: Antihistamine, ACTH, cortisone, hydrocortisone, and anesthetics in snake bite. *Am J Trop Med Hyg, 3:*1083, 1954.

Schwartz, H. J., and Naff, G. B.: The effect of complement depletion by cobra venom factor on delayed hypersensitivity reactions. *Proc Soc Exp Biol Med, 138:*1041, 1971.

Schwick, G., and Dickgiesser, F.: Probleme der Antigen-und Fermentanalyse in Zusammenhang mit der herstellung polyvalenter Schlangengiftseren. In Behringwerke Mitteilungen *Die Giftschlangen der Erde.* Marburg-Lahn, N. G. Elwert, 1963, p. 35.

Scragg, R. F. R., and Szent-Ivany, J. J. H.: Fatalities caused by multiple hornet stings in the territory of Papua and New Guinea. *J Med Ent, 2:*309, 1965.

Sergent, E.: Venin de *Scorpio maurus. Arch Inst Pasteur d'Algerie, 24:*301, 1946.

Shannon, F. A.: Case reports of two Gila monster bites. *Herpetologica, 9:* 125, 1953.

Shannon, F. A.: Report on a fatality due to rattlesnake bite. *Nat Hist Misc, Chicago Acad Sci,* no. 135:7, 1954.

Shapiro, Bert I.: Purification of a toxin from the tentacles of the anemone *Condylactis gigantea. Toxicon, 5:*253, 1968.

Shilkin, K. B.; Chen, B. T. M., and Khoo, O. T.: Rhabdomyolysis caused by hornet venom. *Br Med J, 1:*156, 1972.

Shugihara, H.; Nikai, T.; Moriura; Kamiya, K., and Tanaka, T.: Enzymochemical studies on snake venoms. I. Changes in biological and enzymatic activities of snake venoms on long standing at room temperature. *Jap J Bact, 27*:47, 1972.

Shulman, Sidney: Insect allergy: biochemical and immunological analyses of the allergens. *Progr Allergy, 12*:246, 1968.

Shulov, Aharon: On the poison of scorpions in Israel(II). *Harefuah, 49*: 186, 1955.

Shulov, A.: Biology and ecology of venomous animals in Israel. *Mem Inst Butantan Simp Internac, 33*:93, 1966.

Shulov, A., and Amitai, P.: Observations sur les scorpions: *Orthochirus innesi* E. Sim., 1910 ssp. Negebensis nov. *Arch Inst Pasteur d'Algerie, 38*: 117, 1960.

Shulov, A.; Nitzen, M.; Tzabar, G., and Zerahia, T.: Attempts to immunize scorpions by various antigens. In DeVries, A., and Kochva, E.: *Toxins of Animal and Plant Origin,* New York, Gordon and Breach, vol. III, pp. 927-931, 1973.

Sitprija, V.; Sribhibhadh, R., and Benyajati, C.: Hemodialysis in poisoning by sea snake venom. *Br Med J, 3*:218, 1971.

Skeie, E.: Weeverfish toxin. Extraction methods, toxicity determinations, and stability examinations. *Acta Pathol Micro Scand, 55*:166, 1962a.

Skeie, E.: Weeverfish venom. Some physio-chemical and immunological observations. *Acta Pathol Micro Scand, 56*:229, 1962b.

Skeie, E.: Toxin of the weeverfish *(Trachinus draco)* experimental studies on animals. *Acta Pharm Toxicol, 19*:107, 1962c.

Skeie, E.: Weeverfish stings. *Danish Med Bull, 13*:119, 1966.

Slotta, K., and Fraenkel-Conrat, H.: Two active proteins from rattlesnake venom. *Nature, 142*:213, 1938.

Slotta, K. H., and Vick, J. A.: Identification of the direct lytic factor from cobra venom as cardiotoxin. *Toxicon, 6*:167, 1969.

Smith, Clifton W., and Micks, D. W.: A comparative study of the venom and other components of three species of *Loxosceles. Am J Trop Med Hyg, 17*:651, 1968.

Smith, C. W., and Micks, D. W.: The role of polymorphonuclear leukocytes in the lesion caused by the brown spider, *Loxosceles reclusa. Lab Invest, 22*:90, 1970.

Smith, David S., and Russell, F. E.: Structure of the venom gland of the black widow spider *Latrodectus mactans.* A preliminary light and electron microscope study. In Russell, F. E., and Saunders, P. R.: *Animal Toxins.* Oxford, Pergamon Press, 1967, pp. 1-15.

Smith, F. D.; Miller, N. G.; Carnazzo, S. J., and Eaton, W. B.: Insect bite by *Arilus cristatus,* a North American reduviid. *Arch Dermatol, 77*:324, 1958.

Smith, J. L. B.: A case of poisoning from the stonefish, *Synanceja verrucosa. Copeia,* no. 3, 207, 1951.

Smith, J. L. B.: Two rapid fatalities from stonefish stabs. *Copeia,* No. 3: 249, 1957.

Smith, M. A., and Hindle, E.: Experiments with the venom of *Laticauda, Pseudechis* and *Trimeresurus* species. *Trans Roy Soc Trop Med Hyg, 25:* 115, 1931.

Snyder, C. C.: Tissue excision urged for snakebite. *J A M A, 198:35,* 1966.

Snyder, C. C.; Knowles, R. P.; Pickens, J. E., and Emerson, J. L.: Pathogenesis and treatment of poisonous snake bites. *J Am Vet Med Assoc, 151:* 1635, 1967.

Snyder, C. C.; Straight, R., and Glenn, J.: The snakebitten hand. *Plast Reconstr Surg, 49:275,* 1972.

Southcott, R. V.: Fatal stings to north Queensland bathers. *Med J Aust, 1:* 272, 1952.

Southcott, R. V.: Notes on stings of some venomous Australian fishes. *Med J Aust, 2:722,* 1970.

Spielman, A., and Levi, H. W.: Probable envenomation by *Chiracanthium mildei;* a spider found in houses. *Am J Trop Med Hyg, 19:729,* 1970.

Springer, V. G., and Smith-Vaniz, W. F.: Mimetic relationships involving fishes of the Family Blenniidae. *Smithsonian Contrib Zool, 112:35,* 1972.

Stahnke, Herbert L.: The Arizona scorpion problem. *Ariz Med, 7:23,* 1950.

Stahnke, H. L.: *Scorpions.* Tempe, Ariz., Arizona State University, 35, 1956.

Stahnke, Herbert L.: Some aspects of scorpion behavior. *Bull South Calif Acad Sci, 65:65,* 1966.

Stahnke, Herbert L.: *Diplocentrus spitzeri,* a new Arizona species of scorpion. *Ent News, 81:25,* 1970.

Stahnke, H. L.: Arizona's lethal scorpion. *Ariz Med,* June, 1972.

Stahnke, H. L.; Allen, R. M.; Horan, R. V., and Tenery, J. H.: The treatment of snake bite. *Am J Trop Med Hyg, 6:323,* 1957.

Stahnke, H. L., and McBride, A.: Snakebite and cryotherapy. *J Occup Med, 8:72,* 1966.

Stahnke, H. L., and Johnson, B. D.: *Aphonopelma* tarantula venom. In Russell, F. E. and Saunders, P. R. (Eds.): *Animal Toxins.* Oxford, Pergamon Press, 1967, pp. 35-39.

Stahnke, H. L.; Heffron, W. A., and Lewis, D. L.: Bite of the Gila Monster. *Rocky Mt Med J, 67:25,* 1970.

Steinitz, H.: Observations on *Pterois volitans* and its venom. *Copeia,* No. 2, 158, 1959.

Stillway, L. W., and Lane, C. E.: Phospholipase in the nematocyst toxin of *Physalia physalis. Toxicon, 9:*193, 1971.

Stone, O. J.; Willis, C. J., and Mullins, J. F.: Thiabendazol inhibition of venom necrosis. *J Invest Dermatol, 47:*67, 1966.

Storer, T. I.: *Heloderma* poisoning in man. *Bull Antivenin Inst Am, 5:*12, 1931.

Stringer, J. M.; Kainer, R. A., and Tu, A. T.: Ultrastructural studies of myonecrosis induced by cobra venom in mice. *Toxicol Appl Pharmacol, 18:*442, 1971.

Stringer, J. M.; Kainer, R. A., and Tu, A. T.: Myonecrosis induced by rattlesnake venom. *Am J Pathol, 67:*127, 1972.

Strydom, D. J.: Snake venom toxins. The amino acid sequences of two toxins from *Dendroaspis polylepis* (black mamba) venom. *J Biol Chem, 247:*4029, 1972.

Strydom, A. J., and Botes, D. P.: Snake venom toxins. Purification, properties, and complete amino acid sequences of two toxins from ringhals *(Hemachatus haemachates)* venom. *J Biol Chem, 246:*1341, 1971.

Styblova, Z., and Kornalik, F.: Enzymatic properties of *Heloderma suspectum* venom. *Toxicon, 5:*139, 1967.

Suarez, G. Schenone, H., and Socias, T.: *Loxosceles laeta* venom-partial purification. *Toxicon, 9:*291, 1971.

Sutherland, S. K.: The Sydney funnel-web spider *(Atrax robustus)* 2. Fractionation of the female venom into five distinct components. *Med J Aust, 2:*593, 1972a.

Sutherland, S. K.: The Sydney funnel-web spider *(Atrax robustus)* 3. A review of some clinical records of human envenomation. *Med J Aust, 2:* 643, 1972b.

Sutherland, S. K., and Lane, W. R.: Toxins and mode of envenomation. of the common ringed or blue-banded octopus. *Med J. Aust, 1:*893, 1969.

Sutherland, S. K.; Broad, A. J., and Lane, W. R.: Octopus neurotoxins: low molecular weight non-immunogenic toxins present in the saliva of the blue-ringed octopus. *Toxicon, 8:*249, 1970.

Suzuki, T.: Pharmacologically and biochemically active components of Japanese ophidian venoms. *Mem Inst Butantan Simp Internac, 33:*519, 1966.

Swaroop, S., and Grab, B.: Snakebite mortality in the world. *Bull, World Health Org, 10:*35, 1954.

Swarts, W. B., and Wanamaker, J. F.: Skin blisters caused by vesicant beetles. *J A M A, 131:*594, 1946.

Takahashi, T., and Ohsaka, A.: Purification and characterization of a proteinase in the venom of *Trimeresurus flavoviridis. Biochim Biophys Acta, 198:*293, 1970.

Tallqvist, H., and Osterlund, K.: Huggormsbett. *Nord Med, 68:*1073, 1962.

Tamiya, N., and Arai, H.: Studies on sea snake venoms: Crystallization of erabutoxins a and b from *Laticauda semifasciata* venom. *Biochem J, 99:* 624, 1966.

Taub, Aaron: Ophidian cephalic glands. *J Morphol, 118:*529, 1966.

Taub, Aaron: Comparative histological studies on Duvernoy's gland of colubrid snakes. *Bull Am Mus Nat Hist, 138:*1, 1967.

Taylor, E. H., and Denny, W. F.: Hemolysis, renal failure and death, presumed secondary to bite of brown recluse spider. *South Med J, 59:*1209, 1966.

Theodorides, Jean: The parastological, medical and veterinary importance of Coleoptera. *Acta Tropica, 7:*48, 1950.

Theodorides, J.: Notes sur des coleopteres d'importance medicale. *Med Trop, 14:*318, 1954.

Thompson, T. E., and Bennett, I.: Physalia nematocysts utilized by mollusks for defense. *Science, 166:*1532, 1969.

Tinkham, E. R.: A poison-squirting spider. *Bull U S Army Med Dept, 5:* 361, 1946.

Trethewie, E. R.: Pharmacological effects of the venom of the common octopus *Hapalochlaena maculosa*. *Toxicon, 3:*55, 1965.

Trethewie, E. R.: Detection of snake venom in tissue. *Clin Toxicol, 3:*445, 1970.

Trethewie, E. R., and Rawlinson, P.: Immunological diagnosis of type of snake in snake bite. *Med J Aust, 2:*111, 1967.

Trinca, G. F.: The treatment of snakebite. *Med J Aust, 2:*275, 1963.

Tu, A. T.; Chua, A., and James, G. P.: Peptidase activities of snake venoms. *Comp Biochem Physiol, 15:*517, 1966.

Tu, A. T., and Ganthavorn, S.: Comparison of *Naja naja siamensis* and *Naja naja atra* venoms. *Toxicon, 5:*207, 1968.

Tu, A., and Toom, P. M.: Hydrolysis of peptides by snake venoms of Australia and New Guinea. *Aust J Exp Biol Med Sci, 45:*561, 1967a.

Tu, A. T., and Toom, P. M.: The presence of L-leucyl-βnapthlamide hydrolyzing enzyme in snake venoms. *Experientia, 23:*439, 1967b.

Tu, A. T., and Toom, P. M.: Hydrolysis of peptides by *Crotalidae* and *Viperidae* venoms. *Toxicon, 5:*201, 1968.

Tu, A. T.; Toom, P. M., and Murdock, D. S.: Chemical differences in the venoms of genetically different snakes. In Russell, F. E., and Saunders, P. R. (Eds.): *Animal Toxins*. Oxford, Pergamon Press, 1967, pp. 351-362.

Tu, A. T., and Hong, B. S.: Purification and chemical studies of a toxin from the venom of *Lapemis hardwickii* (Hardwick's sea snake). *J Biol Chem, 246:*2772, 1971.

Tu, A. T., and Toom, P. M.: Isolation and characterization of the toxic component of *Enhydrina schistosa* (common sea snake) venom. *J Biol Chem, 246:*1012, 1971.

Tu, A., and Passey, R. B.: Phospholipase A from sea snake venom and its biological properties. In DeVries, A., and Kochva, E.: *Toxins of Animal and Plant Origin*. 1972, vol. II, pp. 419-436.

Tyler, A.: An auto-antivenin in the Gila Monster and its relation to a concept of natural auto-antibodies. In Buckley, E. E., and Porges, N. (Ed.): *Venoms,* AAAS Publication 44. American Association for Advancement of Science, Washington D. C., 1956, pp. 65-74.

U. S. Navy Bureau of Medicine and Surgery: Poisonous Snakes of the World (rev. ed., 1968). U. S. Government Printing Office, Washington, D. C.

Valle, J.; Picarelli, Z. P., and Prado, J. L.: Histamine content and pharmacological properties of crude extracts from setae of urticating cater-pillars. *Arch Internat Pharmacodyn, 98:*324, 1954.

Van Der Walt, S. J., and Joubert, F. J.: Studies on puff adder *(Bitis arietans)* venom, I. Purification and properties of protease A. *Toxicon, 9:*153, 1971.

Vellard, J.: Resistance de quelques especes animales au venin de serpent. *Compt Rend Acad Sci, 143:*5, 1949.

Vidal, J. C., and Stoppani, A. O.: Isolation and purification of two phospho-lipases A from *Bothrops* venoms. *Arch Biochem Biophys, 145:*543, 1971.

Visser, John: *Poisonous Snakes of Southern Africa.* Cape Town, Howard Timmins, 1966.

Vital Brazil, O.: Pharmacology of crystalline crotoxin. I Toxicity, II Neuro-muscular blocking action. *Mem Inst Butantan Simp Internac, 33:*973, 1966.

Vorse, H.; Seccareccio, P.; Woodruff, K., and Humphrey, G. B.: Dissemi-nated intravascular coagulopathy following fatal brown spider bite. *J Ped, 80:*1035, 1972.

Wagner, F. W., and Prescott, J. M.: A comparative study of proteolytic activities in the venoms of some North American snakes. *Comp Biochem Physiol, 17:*191, 1966.

Wagner, F. W.; Spiekerman, A. M., and Prescott, J. M.: Leucostoma pepti-dase A. *J Biol Chem, 243:*4486, 1968.

Wahlstrom, A.: Purification and characterization of phospholipase A from the venom of *Naja nigricollis. Toxicon, 9:*45, 1971.

Wasuwat, S.: Extract of *Ipomoea pes-caprae* antagonistic to histamine and jellyfish poison. *Nature, 225:*758, 1970.

Waterman, J. A.: Some notes on scorpion poisoning in Trinidad. *Trans Roy Soc Trop Med Hyg, 31:*607, 1938.

Watt, D.: Biochemical studies on the venom from the scorpion, *Centruroides sculpturatus. Toxicon, 2:*171, 1964.

Watt, Dean D., and McIntosh, Max E.: Effects on lethality of toxins in venom from the scorpion *Centruroides sculpturatus. Toxicon, 10:*173, 1972.

Weaver, J. K.: Treatment of rattlesnake bites. Indications for surgery. *Rocky Mt Med J, 67:*31, 1970.

Webb, J. H., and Earnest, F.: Tick paralysis, a case report. *Ohio State Med J, 59:*395, 1963.

Weiss, H. J.; Phillips, L. L.; Hopewell, W. S.; Phillips, G.; Christy, N. P. and Nitti, J. F.: Heparin therapy in a patient bitten by a saw-scaled viper *(Echis carinatus)* a snake whose venom activates prothrombin. *Am J Med, 54:*653, 1973.

Weissman, E., and Shulov, A.: Investigations on the venom of the scorpion *Buthotus judaicus. Arch Inst Pasteur d'Algerie, 37:*202, 1959.

Wellner, D., and Hayes, M. G.: Multiple molecular forms of L-amino acid oxidase. *Ann N Y Acad Sci, 151:*118, 1968.

Wells, M. A., and Hanahan, D. J.: Studies on phospholipase A. Isolation and characterization of two enzymes from *Crotalus adamanteus* venom. *Biochem, 8:*414, 1969.

Welsh, J. H.: Serotonin and related tryptamine derivatives in snake venoms. *Mem Inst Butantan Simp Internac, 33:*509, 1966.

Welsh, J. H.: Acetylcholine in snake venoms. In Russell, F. E., and Saunders, P. R. (Eds.): *Animal Toxins.* Oxford, Pergamon Press, 1967, pp. 363-368.

Welsh, John H., and Batty, Carolyn: 5-Hydroxytryptamine content of some arthropod venoms and venom-containing parts. *Toxicon, 1:*165, 1963.

Wheeling, C. H., and Keegan, H. L.: Effects of scorpion venom on a tarantula. *Toxicon, 10:*305, 1972.

Whittemore, F. W. Jr.; Keegan, H. L., and Borowitz, J. L.: Studies of scorpion antivenins. I. Paraspecificity. *Bull WHO, 25,* 185, 1961.

Whittemore, F. W. Jr.; Keegan, H. L., and Borowitz, J. L.: Venom collection and scorpion colony maintenance. *Bull WHO, 28:*505, 1963.

Wiener, S.: The Australian redbacked spider. *Med J Aust, 2:*331, 1956.

Wiener, Saul: The Sydney funnel-web spider *(A. robustus).* I. Collection of venom and its toxicity in animals. *Med J Aust, 2:*377, 1957.

Wiener, Saul: The production and assay of stone-fish antivenene. *Med J Aust, 2:*715, 1959.

Wiener, S.: Active immunization of man against venom of the Australian tiger snake. *Am J Trop Med Hyg, 9:*284, 1960.

Wiener, Saul: Observations on the venom of the Sydney funnel-web spider *(Atrax robustus). Med J Aust, 2:*693, 1961.

Wiener, Saul: Antigenic and electrophoretic properties of funnel web spider *(Atrax robustus)* venom. In Keegan, H. L., and Macfarlane, W. V.: *Venomous and Poisonous Animals and Noxious Plants of the Pacific Region.* Oxford, Pergamon Press, 1963, pp. 141-151.

Wilde, H.: Anaphylactic shock following bite by a "slow loris" *Nycticebus coucang, Am J Trop Med Hyg, 21:*486, 1972.

Williams, F. E.; Freeman, M., and Kennedy, E.: The bacterial flora of the mouths of Australian venomous snakes in captivity. *Med J Aust, 21:*190, 1934.

Williams, Martin, and Williams, Charles: Collection and toxicity studies of ant venom. *Proc Soc Exptl Biol Med, 116:*161, 1964.

Williams, W. J., and Esnouf, M. R.: The fractionation of Russell's viper *(Vipcra russellii)* venom with special reference to the coagulant protein. *Biochem J, 84:*52, 1962.

Wittle, L. W.; Middlebrook, R. E., and Lane, C. E.: Isolation and partial purification of a toxin from *Millepora alcicornis. Toxicon, 9:*327, 1971.

Wood, J. T.; Hoback, W. W., and Green, T. W.: Treatment of snake venom poisoning with ACTH and cortisone. *Virginia Med Month, 82:*130, 1955.

Woodson, W. D.: Toxicity of *Heloderma* venom. *Herpetologica, 4:*31, 1947.

Wright-Smith, R. J.: A case of fatal stabbing by a stingray. *Med J Aust, 2:* 466, 1945.

Wu, T., and Tinker, D. O.: Phospholipase A2 from *Crotalus atrox* venom. Purification and some properties. *Biochem, 8:*1558, 1969.

Ya, P. M., and Perry, J. F. Jr.: Experimental evaluation of methods for early treatment of snake bite. *Surgery, 47:*975, 1960.

Yaffee, H. S., and Stargardter, F.: Erythema multiforme from *Tedania ignis*. Report of a case and an experimental study of the mechanism of cutaneous irritation from the fire sponge. *Arch Dermatol, 87:*601, 1963.

Yang, C. C.: Crystallization and properties of cobrotoxin from Formosan cobra venom. *J Biol Chem, 240:*1616, 1965.

Yang, C. C.; Yang, H. J., and Huang, J. S.: The amino acid sequence of cobrotoxin. *Biochem Biophys Acta, 188:*65, 1969.

Yarom, Rena: Scorpion venom: a tutorial review of its effects in men and experimental animals. *Clin. Toxicol, 3:*561, 1970.

Zarafonetis, J. D., and Kalas, J. P.: Serotonin, catechol amines, and amine oxidase activity in the venoms of certain reptiles. *Am J Med Sci, 240:* 764, 1960.

Zeller, E. A.: Enzymes of snake venoms and their biological significance. *Adv Enzymol, 8:*459, 1948.

Zerachia, T.; Bergmann, F., and Shulov, A.: Pharmacological activities of the predacious bug *Holotrichus innesi* (Het. Reduviidae). *Toxicon, 10:* 537, 1972.

Zervos, S. G.: La maladie des pecherus d'eponges nus. *Paris Med, 93:*89, 1934.

Ziegler, F. D.; Vasquez-Colon, L.; Elliott, W. B.; Gans, C., and Taub, A.: Studies on energy metabolism following treatment of mitochondria with several snake venoms. In Russell, F. E., and Saunders, P. R. (Eds.): *Animal toxins.* Oxford, Pergamon Press, 1967, pp. 236-243.

Ziprkowski, L.; Hofshi, E., and Tahari, A. S.: Caterpillar dermatitis. *Israel Med J., 18:*26, 1959.

Zlotkin, Eliahu, and Shulov, A.: A simple device for collecting scorpion venom. *Toxicon, 7:*331, 1969.

Zlotkin, E.; Fraenkle, G.; Miranda, F., and Lissitzky, S.: The effect of scorpion venom on blowfly larvae—a new method for the evaluation of scorpion venoms potency. *Toxicon, 9:*1, 1971a.

Zlotkin, E.; Miranda, F.; Kupeyan, C., and Lissitzky, S.: A new toxic protein in the venom of the scorpion *Androctonus australis* Hector. *Toxicon, 9:*9, 1971b.

Zlotkin, E.; Miranda, F., and Lissitzky, S.: Proteins in scorpion venoms toxic to mammals and insects. *Toxicon, 10:*207, 1972a. A factor toxic to crustacean in the venom of the scorpion *Androctonus australis* Hector, 1972b, *Ibid.,* p. 211.

Zwisler, O.: The role of enzymes in the processes responsible for the toxicity of snake venoms. *Mem Inst Butantan Simp Internac, 33:*281, 1966.

INDEX